Challenges in Agro-Climate and Ecosystem

Edited by Muhammad Saifullah, Guillermo Tardio and Slobodan B. Mickovski

Published in London, United Kingdom

Challenges in Agro-Climate and Ecosystem
http://dx.doi.org/10.5772/intechopen.98139
Edited by Muhammad Saifullah, Guillermo Tardio and Slobodan B. Mickovski

Assistant to the Editors: Paola Sangalli

Contributors
Manuel T. Oliveira, Ana A. Oliveira, Yichao Tian, Yongwei Yang, Hüseyin Emre Ilgın, Markku Karjalainen, Dennis Somelar, Festus Kelonye, Godfrey Juma, Valentin Kazandjiev, Veska Georgieva, Petia Malasheva, Dragomir Atanassov, Mohamed M. Blango, Richard A. Cooke, Juana P. Moiwo, Emmanuel Kangoma, Muhammad Saifullah, Muhammad Adnan, Muhammad Arshad, Muhammad Waqas, Asif Mehmood

First published in London, United Kingdom, 2022 by IntechOpen
IntechOpen is the global imprint of INTECHOPEN LIMITED, registered in England and Wales, registration number: 11086078, 5 Princes Gate Court, London, SW7 2QJ, United Kingdom

British Library Cataloguing-in-Publication Data
A catalogue record for this book is available from the British Library

Additional hard and PDF copies can be obtained from orders@intechopen.com

Challenges in Agro-Climate and Ecosystem
Edited by Muhammad Saifullah, Guillermo Tardio and Slobodan B. Mickovski
p. cm.
Print ISBN 978-1-80355-324-5
Online ISBN 978-1-80355-325-2
eBook (PDF) ISBN 978-1-80355-326-9

Meet the editors

Dr. Muhammad Saifullah has research experience in the field of agricultural water resources and climate change. He obtained his Ph.D. from Hohai University, China. He is the recipient of two best researcher awards from the Ministry of Education, China, as well as second place in the Yellow River Hydraulic Research Institute award. He has published international peer-reviewed research articles and investigated different international and national research projects. Dr. Saifullah was an assistant professor at the University of Agriculture Faisalabad and the University of Faisalabad, Pakistan. He was also a researcher at the Institute of International Rivers and Eco-security, Yunnan University, China. He is currently an assistant professor in the Department of Agricultural Engineering, Muhammad Nawaz Shareef University of Agriculture, Pakistan.

Guillermo Tardío is an experienced practitioner, Chartered engineer and researcher at Technical University of Madrid. He is developing and publishing integrated design schemes by enriching traditional engineering design protocols and routines with ecosystem services and eco-engineering approaches. He has specialist expertise in ground bio- and eco-engineering solutions, erosion protection and, Nature Based Solutions (NBS) design. He has a broad experience in river restoration and slope stabilization projects. Guillermo holds a Ph.D. and Msc degrees in Forestry Engineering from the Technical University of Madrid (Spain). He has authored a patented system for erosion control in gullies.

Prof. Slobodan B. Mickovski is an experienced academic with a civil and geotechnical engineering background who has worked in sectors including rail, highways, flood defense, marine, residential developments, car parks, erosion protection, and forestry. He has specialist expertise in ground bio- and eco-engineering solutions and erosion protection through sustainable use of vegetation including vegetated natural or humanmade slopes, nature-based solutions, and natural hazard remediation. He has led and participated in several national and international research projects including Knowledge Network Partnerships, Erasmus, and Horizon2020 projects. He teaches several subjects across environmental civil engineering, environmental management, and renewable energy curricula. He supervises multi-disciplinary master's and Ph.D. research on the boundary between engineering, natural sciences, and social sciences.

Contents

Preface IX

Section 1
Agro-Climate Challenges 1

Chapter 1 3
Temperature Based Agrometeorology Indices Variability in South Punjab, Pakistan
by Muhammad Saifullah, Muhammad Adnan, Muhammad Arshad, Muhammad Waqas and Asif Mehmood

Chapter 2 17
Evapotranspiration and Drought in Different Agricultural Zones of Bulgaria
by Valentin Kazandjiev, Veska Georgieva, Petia Malasheva and Dragomir Atanassov

Chapter 3 35
Modeling the Effect of Climate Change on Water Stored above a Micro-Dam in an Inland Valley Swamp in Sierra Leone, Using SWAT
by Mohamed M. Blango, Richard A. Cooke, Juana P. Moiwo and Emmanuel Kangoma

Chapter 4 51
Exotic Vs. Autochthonous Grapevine Varieties – A Case Study on Global Warming in Northeastern Portugal
by Manuel T. Oliveira and Ana A. Oliveira

Chapter 5 69
Climate Change and Food System in Kenya: Challenges and Opportunities
by Festus Kelonye and Godfrey Juma

Section 2
Ecosystem 79

Chapter 6 81
Spatio-Temporal Variation of Ecosystem Services and Its Trade-off Relationships in Southwest Guangxi
by Yichao Tian and Yongwei Yang

Chapter 7 99
Wooden Extra Stories in Concrete Block of Flats in Finland as an Ecologically Sensitive Engineering Solution
by Markku Karjalainen, Hüseyin Emre Ilgın and Dennis Somelar

Preface

The world's population is increasing, which is increasing the pressure to increase crop yields and produce more food to ensure global food security. This can only be accomplished through the use of science and technology in agriculture to improve the crop varieties capable of producing high yields and management of resources such as water and soil. The impact of increasing population on agriculture is much more complex than the proportional increase in greenhouse gas emissions to the environment. Rising temperatures are changing natural environments and ecosystems across the globe.

Rapid industrialization and urbanization as well as increasing population are degrading the ecosystem and warming the climate. These effects are more serious in developing countries where climate change and environmental degradation are becoming major challenges for crop production. The book addresses the challenges of water stress, temperature stress, increasing evapotranspiration, drought stress, and agrometeorology variability on crop production, food sustainability, and the ecosystem.

Dr. Muhammad Saifullah
Assistant Professor,
Department of Agricultural Engineering,
MNS University of Agriculture Multan,
Multan, Punjab, Pakistan

Guillermo Tardio
Technical University of Madrid (UPM),
Technical Director of ODPE Ltd.,
President of the Spanish Association of Soil Bioengineering,
Member of EFIB (European Federation of Soil Bioengineering),
Madrid, Spain

Slobodan B. Mickovski
BEAM Centre,
Glasgow Caledonian University,
Glasgow, Scotland, UK

Section 1

Agro-Climate Challenges

Chapter 1

Temperature Based Agrometeorology Indices Variability in South Punjab, Pakistan

Muhammad Saifullah, Muhammad Adnan, Muhammad Arshad, Muhammad Waqas and Asif Mehmood

Abstract

Climate change has a major impact on crop yield all over the world. Pakistan is one of the major affected countries by climate change. The agrometeorology indices were determined for the South Punjab region, which is a hot spot for climate change and food security. This region is rich in agriculture, but crop yield relationship is estimated with agrometeorology indices (AMI). Temperature stress (33°C), average diurnal temperature range (12°C), Average accumulative growing degree days (1303°C), phototemperature (27°C) and nyctotemperature (21°C) indices were determined for Multan. The variation in diurnal temperature was found at 0.39 for Bahawalpur region and similar variation was observed in growing degree days, which is 0.11 more than the diurnal temperature range. The extreme of these indices which influence the crop yield was found in May and June. The cropping period from sowing to harvest varied due to climate change and cause to decrease in the yield of the crop. The indices are regarded as crop performance indicators. So, policymakers and agricultural scientists should take necessary measures to mitigate such kinds of challenges.

Keywords: indices, South Punjab, temperature, meteorology, indices, Pakistan

1. Introduction

During the last decade, climate change has raised an international attention due to concerns of negative impacts on ecosystems, agriculture, water supply and management, human welfare and regional political stability [1, 2]. The global situation is well described by the IPCC assessment report [3], which in particular cites global mean temperatures as having risen approximately by 0.74°C in the last 100 years. The majority of this warming has occurred since 1950, most likely due to increasing greenhouse gas concentrations to unprecedented levels in recent history [4]. Global climate models (GCMs) predict temperature increases of 1.1–6.4°C over 1990–2100 [3]. Due to the great influence of temperature on evapotranspiration and precipitation for soil water availability and drought events, climate change might have an immediate, direct effect on vegetation and crop productivity and, therefore, on the related net income for farmers [1, 5, 6].

Climate change is one of the most prominent global environmental issues. During the period from 1885 to 2012, the mean global temperature has increased by 0.85°C and is predicted to increase further by 1.6–5.8°C by the end of 21st century [3]. Climate is one of the most critical limiting factors for agricultural production: frost risk during the growing period and low and irregular precipitation with high risks of drought during the cultivating period are common problems in agriculture. In recent years a change in climate has been documented in many locations throughout the world. Increasing rainfall trends were reported in Argentina [7], Australia and New Zealand [8, 9]. The minimum temperature increased almost everywhere. The maximum and mean temperature increased in northern and central Europe, over the Bulgaria [10], Canada [11], Australia, New Zealand [8], India [12]. These results support the assumptions [13] that mid-latitude regions such as the mid-western USA, southern Europe and Asia are becoming warmer and drier, whereas the lower latitudes are becoming warmer and wetter.

Developing countries are more vulnerable to such changes as they have limited resources to cope up with the disasters and agriculture plays dominant role in their national economy [14]. The northern part of Indian sub-continent has been placed under high risk zone for heat stress risks in view of future climate change scenarios [15]. Under such conditions, the sustainability of natural resources and food security for the increasing population in the region is at risk. An increase, even moderate in global temperature is expected to result in a change in frequency of extreme weather events like drought, heavy rainfall and storms [16]. Small changes in precipitation mean result in a relatively high increase in the probability of precipitation extremes [17, 18]. The same effect has been demonstrated for temperature changes [19]. Wagner [20] suggest that the frequency of extreme events is relatively more dependent on a climate change as compared to mean values of climatic parameters.

Ali et al., [21] also determined the future trends of temperature in different regions of Pakistan. Nadeem et al. [22] found the results using different climate models in south Punjab, Pakistan. The study period divided into base line period (1980–2018) and found the increase in average temperature 0.94°C for this period. According to Global Climate Models increased in temperature 3–5°C up to 2050. The yield of different crops decreased and sowing 18–25 days earlier as adaptation startgey to cope this climate change challenge. Punjab is also experiencing large fluctuations in temperature and precipitation patterns every year leading to large oscillations in agricultural productivity in the region. The unpredictable weather conditions have already started to diverse impact on productivity of wheat crop during last 5–6 years. To manage this alarming situation, there is a need to analyze the spatiotemporal variability of climatic conditions at regional scale so that viable mitigation/adaptation strategies could be developed and implemented on regional basis. Keeping in this view, the spatiotemporal climatic variability during kharif and rabi seasons has been studied for three different agroclimatic regions of South Punjab by using statistical approaches.

2. Materials and methods

Daily minimum (T_{min}) and maximum (T_{max}) temperature from two stations in South Punjab (SP), Pakistan (**Table 1**, **Figure 1**) with the longest and complete records were used. First, monthly variations were identified in temperature series

Station	Average	Median	Standard deviation	Variance	skewness	Co-efficient of variation
			Maximum Temperature			
Multan	31.5	32.5	8.27	68.32	−0.31	0.26
Bahawalpur	32.1	33.8	8.08	65.26	−0.44	0.25
			Minimum Temperature			
Multan	19.6	20.80	8.83	77.99	−0.26	0.45
Bahawalpur	19.10	20.20	8.56	73.32	−0.44	0.45

Table 1.
Statistics of minimum and maximum temperature.

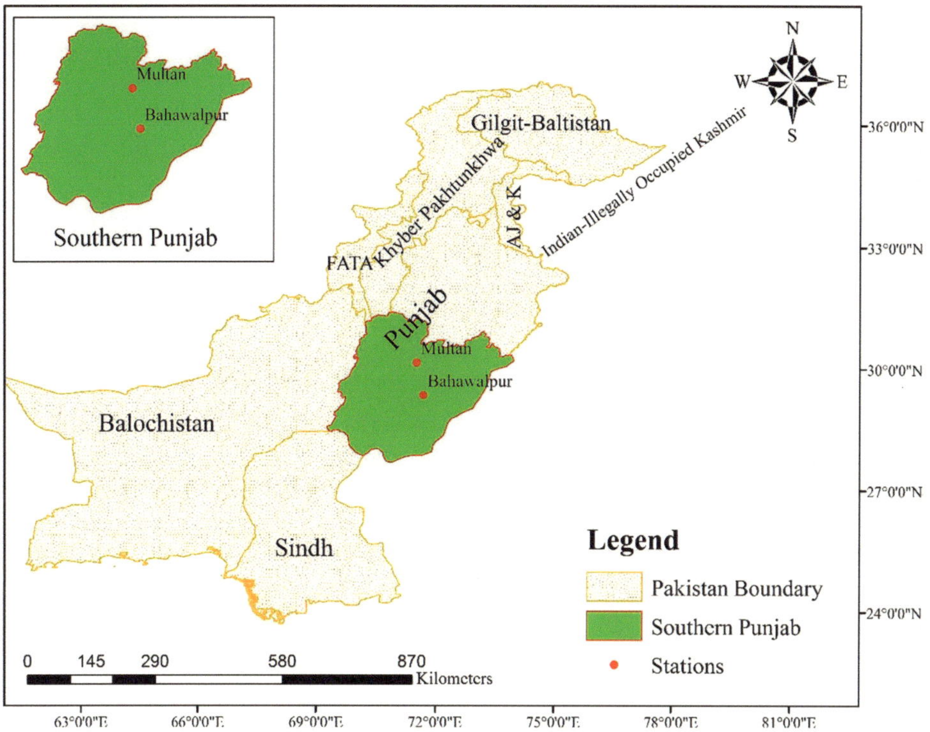

Figure 1.
Location of study area.

considering the study period. Multan and Bahawalpur stations were chosen for the time of 2012–2016. Both divisions comprised a major part of South Punjab. The region is considered very important regarding the agriculture production. Wheat and Cotton crops were dominated for the South Punjab region (**Table 1**).

3. Methods

Determination of following temperature base agro metrological indices has carried out. Crops growth when the daily T_{mean} is above a given temperature threshold.

Indices	Abbreviation	Description	Unit
Diurnal Temperature Range	DTR	Difference between daily maximum and minimum temperature	°C
Growing Degree Days	GDD	Accumulated number for temperature degrees above 30°C	°C
Temperature stress >30°C	T_{30}	Average Temperature threshold above 30°C	°C
Nyctotemperature	T_n	Mean temperature during night when light levels are limited	°C
Phototemperature	T_b	Effective light temperature	°C

Table 2.
Temperature base agro-meteorological indices.

Different temperature base indices are used to quantify that climate change process and its effect on crop production area and yield. The temperature base agro meteorological indices are given in **Table 2**.

The maximum, minimum and average temperature of South Punjab is extreme in month of May and June (**Figure 2**). Month of July and August is identified lower as compared to extreme months due to moon soon spell.

To explore the variations of temperature during the months, the Diurnal Temperature Range (DTR) [23] is determined:

$$DTR\left({}^{\circ}C\right) = T_{\max} - T_{\min} \tag{1}$$

Chilling temperature is most important during the dormancy period. It also plays key role in the growth of plants in spring season. Deciduous fruit trees demanded the accumulating chilling temperature in temperate environment.

When the daily temperature exceeds from threshold temperature. Threshold temperature can vary crop as well as species and stages of crop. Quantify the thermal

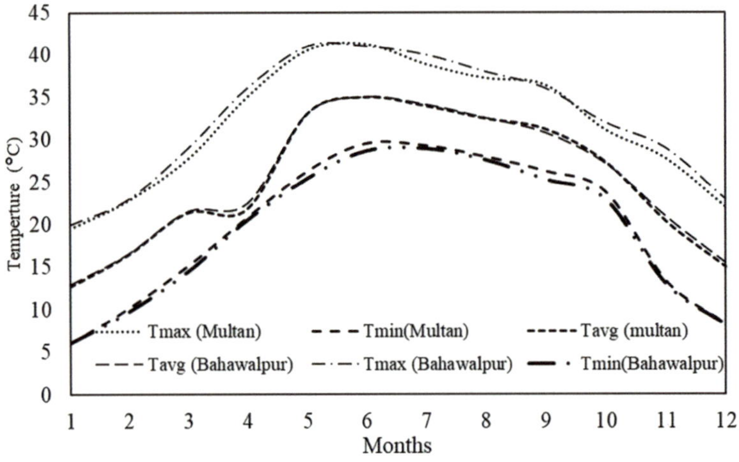

Figure 2.
Average, maximum and minimum temperature of Multan and Bahawalpur.

time and growing degree day used to estimate the process. It also helps for selection of crops and hybrid varieties to achieve the maximum growth at the time maturity and yield. The growing degree day is determined [24, 25]:

$$GDD = \sum_{1}^{n} (T_{aver} - T_b) \quad (2)$$

Where, n is the number of days in each month, T_{mean} is the daily average temperature computed using daily T_{min} and T_{max}, T_b is threshold temperature for the crop growth.

After decided the analyses of growing degree days, the heat unit is most important indices for the crops. The cumulative heat unit is calculated by summing the daily mean temperature above the base temperature [26]: The effective light temperature estimated from the given equation:

$$T_P = T_{max} - 0.25DTR \quad (3)$$

This index is computed for the different stages and significance for mean temperature at daytime. It is named as Phototemperature index [27]. There is also other index to compute the effect of mean temperature at full nighttime:

$$T_n = T_{min} + 0.25DTR \quad (4)$$

It is named as nyctotemperature index for given crop during the different stages. There are also identify the maximum and minimum temperature of the year highlighted for this region and alarm all the stakeholder regarding the climate change warming.

4. Results and discussions

The results reveled that different agro-meteorology indices varied whole the time period for this region. The variations diurnal temperature range is more as highlighted from coefficient of variations. These variations of Diurnal temperature range observed 5% higher in Multan as compared to Bahawalpur. Growing degree day of Multan 2% higher than Bahawalpur. Temperature stress play important role for the crop growth and phenology at different stages, South Punjab region have rich in agriculture and playing key role textile industry. Textile industry raw materials produced from this region because it cottons dominant region.

The temperature above the 30°C, this index found to be similar for both stations. Phototemperature index Multan is higher 5% than Bahawalpur (**Table 3**). The variations in Nyctotemperature of Bahawalpur index lower 2% higher than Multan. The interanual variations in diuneral temperture range found extreme in month of May and June. The lowest DTR highlighted in month of October. The higher variability indicated from diuneral temperture range during the different months of 2012–2016 (**Figure 3**).

Heat stress found to be maximum in month of June for period 2012–2016 (**Figure 4**). Temperature stress of Multan was 1.5% higher than Bahawalpur for the month of June. The variations in heat stress found to be more in Multan as compared

Indices	Average	Median	Cv
Multan			
DLR	12	12.4	0.41
GDD	1303	1331	0.6
T_{30}	33.7	33.4	0.1
T_n	21.4	23.30	0.45
T_p	27	28.79	0.38
Bahawalpur			
DLR	13	13.5	0.39
GDD	1277	1460	0.5
T_{30}	33.5	33.5	0.10
T_n	21.2	23.10	0.44
T_p	27.4	29.84	0.36

Table 3.
Descriptive statistics of agrometeorological indices of South Punjab.

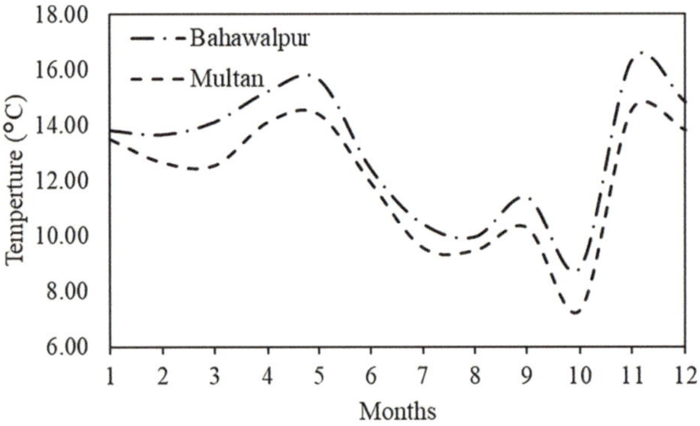

Figure 3.
Diurnal temperature range for study period.

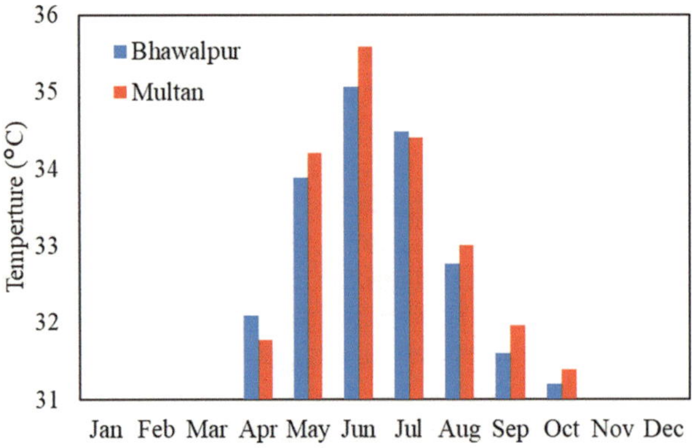

Figure 4.
Temperature stress above threshold for study period.

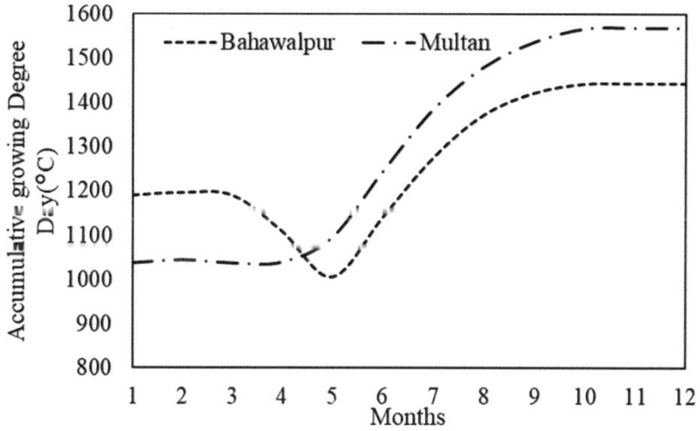

Figure 5.
Growing degree days indices of study area.

to Bahawalpur. For month of April, heat stress was 32.10°C in Bahawalpur, which was 0.7°C more from Multan. Maximum variance in heat stress in month of May were noticed for Multan. Similar, minimum variance determined in month of September for Bahwalpur. The variation of heat stress determined in month of May and June for both stations. These results consistent with the findings of [28].

Accumulative growing degree days increased after the month of June. Multan found to be higher from June to December but lower during the month January to May (**Figure 5**). It can be found the matching point of both stations occurred in month of May. The accumulative growing degree day become constant in month of November and December for both stations, but Multan found to be 8% more than Bahwalpur. Minimum growing degree day ware in month of May for Bahawalpur and month January as well as March for the stations of Multan [25].

Phototemperature and Nyctotemperature index were determined for this region. May and June are critical month for temperature-based agriculture index. August and September have decreased variations due to moon soon effect. Winter seasons found to be lower regarding the temperature based agro-meteorology index. The extreme event of both index Bahawalpur is June with different amount. For Multan, phototemperature and nyctotemperture index found to maximum in month June with 18% variations. Both indices observed minimum in month of January, which is lower 71% and 57% respectively. Overall, month of May and June found to be maximum and minimum in month December and January respectively as shown in **Figures 6** and 7.

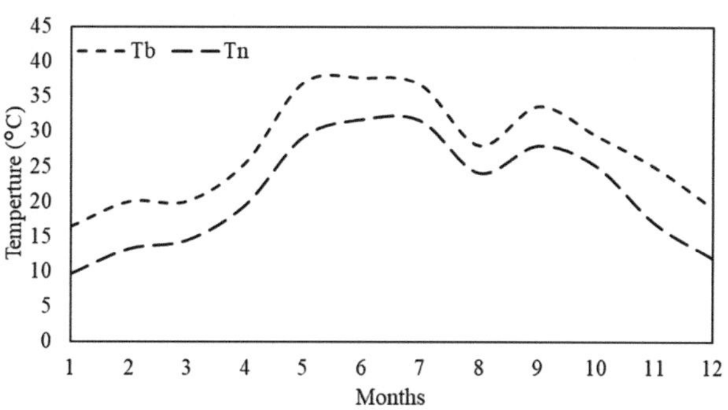

Figure 6.
Phototemperature and Nyctotemperature index for Bahawalpur city.

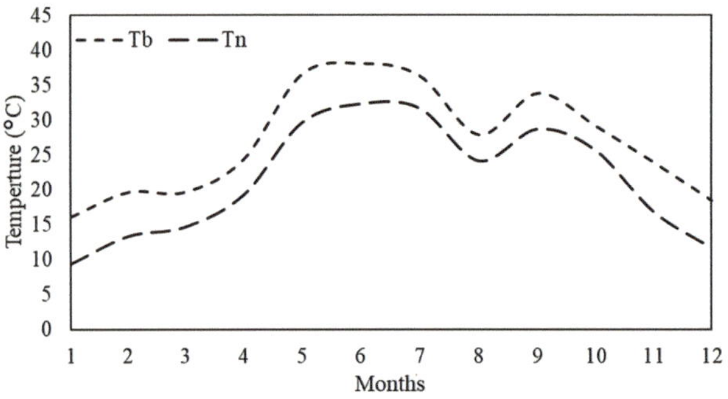

Figure 7.
Phototemperature and Nyctotemperature index for Multan city.

5. Discussions

In this study, we analyze the different agro-meteorology index, which is based on maximum and minimum temperature. Over all these indices indicate global warming decrease the average yield of wheat crop and cotton yield increase in this region. Brar and Singh [29] determined the relationship of agro-meteorology indices (AMI) and cotton yield, the experiments was conducted in north western part of India. The water scarcity was major challenge for cotton production. The crop yield were found increased with higher AMI with heavy pre-sowing irrigation and early sown crop at irrigation with four weeks. The duration to achieve the different phenological stages of wheat with different irrigation treatments for significant higher yield and growing degree days for Rajasthan, India [30]. The uneven trends in temperature indices in Serbia were found the period of 1961–2010. The temperature for growing seasons were complex pattern. The examined temperature indices were found the cooling tendency during the growing seasons and a warming tendency during the dormancy. The warming tendency for the period of 1981–2010 was detected in both seasons with similar magnitude [31]. Jan et al. [32] examined the variability in biological yield of crop with growing degree days with regression model. Fernández-Long et al. [23] investigated the trends in Argentina and agro-meteorology indices identified the potential effect of weather on crop yield and arise a challenge for management decisions. Lalić et al. [1] also highlighted the relation between agro-metrological conditions and crop yield through modeling in different countries. Capra et al. [5] determined the maximum, minimum and mean temperature variability in Italy and found to be very complex behavior with crop growth as well as production. Choudhary et al. [33] also determined the agro-meteorology indices for India and found the relation with crop. These indices identified the crop performance as indicator.

The crop yield and indices found to be closely related. The extreme temperature influence on grain yield and growth of the crop. Due to rapid GDDs shorten the span life of crop and reduce the yield. From this study, the crop yield (**Figures 8** and **9**) is reducing to due to increase in temperature as well as temperature base agro-meteorology indices, which is indicator of challenge of wheat crop in southern part of Punjab and other hand cotton crop is arising. These results were found to be consist with [34, 35]. The temperature stress, phototemperature and nyctotemperture index were also same behavior for crops as findings. Shaheen et al. [36] investigated the GDDs and estimated the shorten the length of growing seasons of crop, which ultimately affect the crop yield, which is consistent with current study. You et al. [37] determined the crop yield and temperature increase relationship for China. The increase

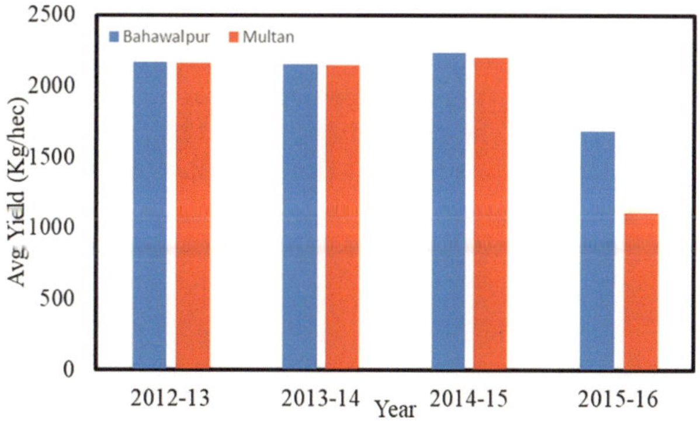

Figure 8.
Wheat yield for the study period.

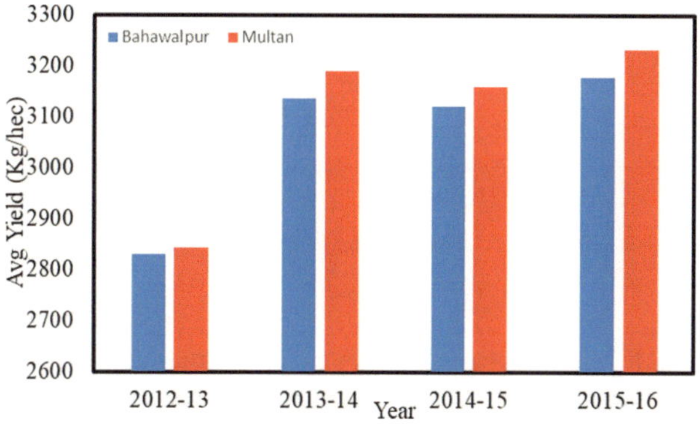

Figure 9.
Cotton yield for study period.

in temperature 1°C reduce the yield 3–10% for mainland of China. The wheat yield decreased as compared to initial year, while cotton yield increase for the same period for region of South Punjab.

6. Conclusion

From this study, it can be concluded that there is a necessity of management and applications of these indices to allocate inter-comparison of the significant results and to improve the understanding. The threshold temperature 30°C identified through this study. It will be helpful for developing whether calendar of wheat, crop insurance product and breeding temperature stress resistant genotype of crops for South Punjab, Pakistan. Month of May and June are more important for this region due to extreme values of temperature base index occurs in these months. These agro-meteorology indices demand to investigate it local and regional basis. There is also necessary to investigate at different time scale i.e., short, and long term. The extreme temperature impact on crop production and management should be locally assessed. Due to extreme temperature base agro-meteorology index in South Punjab, extreme weather is becoming great challenge for farmer community, which effect the livelihood of this region. It creates food security challenge for this region. Due to extreme variations in agro-meteorology stresses, the new varities of crop is needed to introduce which are suitable with current variations in climate. Due to increase in temperature, there is also needed to cope the water scarcity challenge in this region. Therefore, economically water cheap technique should be introduced in this region. The policy maker and scientific community need to take measures to cope with this challenge.

Author details

Muhammad Saifullah[1*], Muhammad Adnan[2], Muhammad Arshad[3,4], Muhammad Waqas[1] and Asif Mehmood[5]

1 Department of Agricultural Engineering, Muhammad Nawaz Shareef University of Agriculture, Multan, Pakistan

2 Institute of International Rivers and Eco-security, Yunnan University, Kunming, Yunnan, China

3 Nanjing University of Information Science and Technology, School of Atmospheric Sciences, Nanjing, China

4 Pakistan Meteorological Department, Islamabad, Pakistan

5 Key Laboratory of Water Cycle and Related Land Surface Processes, Institute of Geographic Sciences and Natural Resources Research (IGSNRR), Chinese Academy of Sciences (CAS), Beijing, China

*Address all correspondence to: saif_2146@yahoo.com

References

[1] Lalić B, Eitzinger J, Thaler S, Vučetić V, Nejedlik P, Eckersten H, et al. Can agrometeorological indices of adverse weather conditions help to improve yield prediction by crop models? Atmosphere. 2014;**5**(4):1020-1041. DOI: 10.3390/atmos5041020

[2] Saifullah M, Adnan M, Zaman M, Wałęga A, Liu S, Khan MI, et al. Hydrological response of the Kunhar River basin in Pakistan to climate change and anthropogenic impactson runoff characteristics. Water. 2021;**13**(22):3163

[3] Change IC. Synthesis Report. Contribution of working groups I. II and III to the fifth assessment report of the intergovernmental panel on climate change. 2014;**151**:(10.1017)

[4] Chavas DR, Cesar Izaurralde R, Thomson AM, Gao X. Long-term climate change impacts on agricultural productivity in eastern China. Agricultural and Forest Meteorology. 2009;**149**(6-7):1118-1128

[5] Capra A, Consoli S, Scicolone B. Long-term climatic variability in Calabria and effects on drought and agrometeorological parameters. Water Resources Management. 2013;**27**(2):601-617. DOI: 10.1007/s11269-012-0204-0

[6] Moonen AC, Ercoli L, Mariotti M, Masoni A. Climate change in Italy indicated by agrometeorological indices over 122 years. Agricultural and Forest Meteorology. 2002;**111**(1):13-27. DOI: 10.1016/S0168-1923(02)00012-6

[7] Viglizzo EF, Roberto ZE, Filippin MC, Pordomingo AJ. Climate variability and Agroecological change in the central pampas of Argentina. Agriculture, Ecosystems & Environment. 1995;**55**(1):7-16

[8] Plummer N, Salinger MJ, Nicholls N, Suppiah R, Hennessy KJ, Leighton RM, et al. Changes in climate extremes over the Australian region and New Zealand during the twentieth century. In: Weather and Climate Extremes. Dordrecht: Springer; 1999. pp. 183-202

[9] Suppiah R, Hennessy KJ. Trends in Total rainfall, heavy rain events and number of dry days in Australia, 1910-1990. International Journal of Climatology: A Journal of the Royal Meteorological Society. 1998;**18**(10):1141-1164

[10] Popova Z, Ivanova M, Pereira L, Alexandrov V, Kercheva M, Doneva K, et al. Droughts and climate change in Bulgaria: Assessing maize crop risk and irrigation requirements in relation to soil and climate region. Bulgarian Journal of Agricultural Science. 2015;**21**(1):35-53

[11] Zhang X, Vincent LA, Hogg WD, Niitsoo A. Temperature and precipitation trends in Canada during the 20th century. Atmosphere-Ocean. 2000;**38**(3):395-429. DOI: 10.1080/07055900.2000.9649654

[12] Duhan D, Pandey A, Gahalaut KPS, Pandey RP. Spatial and temporal variability in maximum, minimum and mean air temperatures at Madhya Pradesh in Central India. Comptes Rendus - Geoscience. 2013;**345**(1):3-21. DOI: 10.1016/j.crte.2012.10.016

[13] Smith B, Ludlow L, Brklacich M. Implications of global climatic warming for agriculture: A review of appraisal (reviews and analyses). Journal of Environmental Quality (EUA). 1988;**17**(4):519-527

[14] Majumder MS, Santillana M, Mekaru SR, McGinnis DP, Khan K, Brownstein JS. Utilizing nontraditional data sources for near real-time estimation of transmission dynamics during the 2015-2016 Colombian Zika virus disease outbreak. JMIR Public Health and Surveillance. 2016;**2**(1):e5814

[15] Nachtergaele F, van Velthuizen H, van Engelen V, Fischer G, Jones A, Montanarella L, et al. Harmonized World Soil Database (Version 1.2). FAO, Rome, Italy: IIASA, Laxenburg, Austria; 2012. pp. 1-50

[16] Balling Jr, Robert C, and Sherwood B Idso. "Effects of greenhouse warming on maximum summer temperatures." Agricultural and Forest Meteorology. 1990;**53**(1-2):143-147

[17] Groisman PY, Karl TR, Easterling DR, Knight RW. Changes in the probability of heavy precipitation: Important indicators of climatic change. In: Jamason PF, Hennessy KJ, Suppiah R, Page CM, Wibig J, Fortuniak K, editors. Weather and Climate Extremes. Dordrecht: Springer; 1999. pp. 243-283

[18] Waggoner A, De Biasio R, Conrad P, Bright GR, Ernst L, Ryan K, et al. Multiple spectral parameter imaging. Methods in Cell Biology. 1989;**30**:449-478

[19] Folland GB. Real Analysis: Modern Techniques and their Applications. Hoboken, New Jersey: John Wiley & Sons; 1999;**40**

[20] Wagner R. Database and expert systems applications. In: 7th International Conference, DEXA'96, Zurich, Switzerland, September 9-13, 1996. Proceedings. Berlin/Heidelberg: Springer Science & Business Media; 1996;**7**

[21] Ali S, Eum HI, Cho J, Dan L, Khan F, Dairaku K, et al. Assessment of climate extremes in future projections downscaled by multiple statistical downscaling methods over Pakistan. Atmospheric Research. 2019;**222**:114-133. DOI: 10.1016/j.atmosres.2019.02.009

[22] Nadeem M, Nazer M, Ghulam K, Fatima Z, Iqbal P, Ahmed M, et al. Application of CSM-CANEGRO model for climate change impact assessment and adaptation for sugarcane in semi-arid environment of southern Punjab, Pakistan. International Journal of Plant Production. Apr 2022. pp. 1-24. DOI: 10.1007/s42106-022-00192-6

[23] Fernández-Long ME, Müller GV, Beltrán-Przekurat A, Scarpati OE. Long-term and recent changes in temperature-based Agroclimatic indices in Argentina. International Journal of Climatology. 2013;**33**(7):1673-1686. DOI: 10.1002/joc.3541

[24] Kash Yapi A, Das HP. Requirement of heat unit and agrometeorological indices in selected wheat growing zones. Mausam. 1999;**january**:63-70

[25] Mcmaster GS, Wilhelm WW. Growing degree-days: One equation, two interpretations Gregory. Agricultural and Forest Meteorology. 1997;**87**(1):291-300

[26] Sattar A, Singh G, Singh SV, Mahesh Kumar P, Kumar V, Bal SK. Evaluating temperature thresholds and optimizing sowing dates of wheat in Bihar. Journal of Agrometeorology. 2020;**22**(2):158-164

[27] Zhu X, Guo R, Liu T, Kun X. Crop yield prediction based on agrometeorological indexes and remote sensing data. Remote Sensing. 2021;**13**(10):2016. DOI: 10.3390/rs13102016

[28] Solanki NS, Devi SS, Chouhan BS, Nai G. Agrometeorological indices, heat use efficiency and productivity of wheat (Triticum Aestivum) as influenced by

dates of sowing and irrigation. Journal of Pharmacognosy and Phytochemistry. 2017;**6**(3):176-180

[29] Brar HS, Singh P. Relationship of agro-meteorological indices with cotton yield under varied pre-sowing irrigation levels, sowing dates and time of first irrigation in North-Western India. Communications in Soil Science and Plant Analysis. 2022;**53**(2):170-179. DOI: 10.1080/00103624.2021.1984513

[30] Prajapat AL, Saxena R, Sharma M, Mandeewal RL, Lal B, Didal B. Growing degree days requirement and yield of wheat cultivars as influenced by irrigation scheduling and time of sowing. International Journal of Plant & Soil Science. 2022;**February**:28-35. DOI: 10.9734/ijpss/2022/v34i330843

[31] Ruml M, Gregorić E, Matović G, Radovanović S, Počuča V. Uneven trends of temperature indices during the growing season and dormancy in Serbia. Theoretical and Applied Climatology. 2022;**147**(3-4):1277-1295. DOI: 10.1007/s00704-021-03859-8

[32] Jan B, Anwar Bhat M, Bhat TA, Yaqoob M, Aijaz Nazir M, Bhat A, et al. Evaluation of seedling age and nutrient sources on phenology, yield and agrometeorological indices for sweet corn (Zea Mays Saccharata L.). Saudi Journal of Biological Sciences. 2022;**29**(2):735-742. DOI: 10.1016/j.sjbs.2021.10.010

[33] Choudhary D, Raj Singh CS, Dagar AK, Singh S. Temperature based agrometeorological indices for Indian mustard under different growing environments in Western Haryana, India. International Journal of Current Microbiology and Applied Sciences. 2018;**7**(1):1025-1035. DOI: 10.20546/ijcmas.2018.701.123

[34] Kingra PK, Setia R, Singh S, Kaur J, Kaur S, Singh SP, et al. Climatic variability and its characterisation over Punjab, India. Journal of Agrometeorology. 2017;**19**(3):246-250

[35] Premdeep ML, Niwas R, Singh M, Kumar S. Agrometeorological indices in relation to phenology of wheat. Journal of Pharmacognosy and Phytochemistry. 2019;**8**(1):2500-2503

[36] Shaheen N, Jahandad A, Goheer MA, Ahmad Q-u-A. Future changes in growing degree days of wheat crop in Pakistan as simulated in CORDEX South Asia experiments. APN Science Bulletin. 2020;**10**(1):82-89. DOI: 10.30852/sb.2020.1221

[37] You L, Rosegrant MW, Wood S, Sun D. Impact of growing season temperature on wheat productivity in China. Agricultural and Forest Meteorology. 2009;**149**(6-7):1009-1014. DOI: 10.1016/j.agrformet.2008.12.004

Chapter 2

Evapotranspiration and Drought in Different Agricultural Zones of Bulgaria

Valentin Kazandjiev, Veska Georgieva, Petia Malasheva and Dragomir Atanassov

Abstract

Bulgaria is located in an area with insufficient humidification and is characterized by periods of the drought of varying duration and intensity. From the last 15 to 20 years, the limiting factor in agro-meteorological conditions was the drought. Agro-meteorological drought consists of the depletion of available soil water reserves in the root zone. Ultimately, the results of droughts affect the size of yields and the quality of production. The consequences of this extreme condition from a meteorological and agro-meteorological point-of-view phenomenon can be mitigated only by expanding the irrigated areas. The aim of this work was to present the tendencies in the change of the potential evapotranspiration during the studied period 1986–2015 as a result of the climate changes and also the tendencies of the conditions for occurrence of agricultural drought in the future, and also to propose an approach where using certain indicators controls the process of accumulation and consumption of water in the soil. Such approach could find application in adapting agriculture to climate change and the updation of agro-environmental zoning relevant to climatic changes.

Keywords: potential evapotranspiration (ETP), soil moisture index (SMI), drought index (AI), annual and seasonal amounts of precipitations and temperatures (Σt and Σr)

1. Introduction

Droughts have a significant impact on agriculture to limit crop productivity and lead to reduced yields and have caused significant economic losses in a number of areas in Europe and the world. According to research by the Institute for Environment and Sustainable Development from the Joint Research Center in Ispra, Italy, drought is one of the biggest related to meteorological disasters. Continuing over months or years, it can affect large areas and can have serious environmental, social and economic impacts. These impacts depend on the duration, severity, and spatial extent of the absence of precipitation, but also on the environment and the socio-economic vulnerability of the affected regions. Europe has rich freshwater resources, but there is a strong regional imbalance across the continent. Water scarcity, for example, is a significant problem in many European regions, particularly in semi-desert and continental climate zones.

A recent study conducted jointly by the European Commission and the Member States estimates the cost of droughts in Europe over the last 30 years at least € 100 billion. The same study estimated the economic damage from drought and heatwaves in 2003 in Central and Western Europe at more than € 12 billion. Other examples of this are the drought that developed in late 2004 in southern Portugal and Spain; the 2006 spring drought in France and the southeast of the United Kingdom and the spring 2011 drought in the enlarged parts of Western Europe, with severe economic consequences, mainly in the agricultural sector. In addition, April 2007 was the driest April according to the meteorological services of Germany, the Netherlands, and Austria, and November 2011 was the driest November in history for large parts of Europe, and Bulgaria. More facts - in the last 5 years in Bulgaria and many countries in the region there is a permanent summer drought, which usually begins in the second half of July and turns into an autumn drought by October and sometimes November.

Climate change, according to the Intergovernmental Panel on Climate Change for Europe, shows significant changes in the water balance across Europe, with an increased likelihood of summer droughts in the Mediterranean, as well as in Central and South-Eastern Europe. However, changes in the annual distribution of precipitation, as well as in energy and water balances, are likely to occur in other regions of Europe, leading to an increased likelihood of declining water levels and an increased likelihood of extreme weather events.

The European Commission published a Communication on "Meeting the Challenges of Water Scarcity and Droughts in the European Union" in December 2007, requesting a wider range of activities to adapt and mitigate the effects of drought and the changing climate in Europe. The measures requested include the development of a European Drought Observatory (EDO), the provision of consistent and timely drought information from the continental to the regional and local scales.

In order to detect, monitor, and forecast droughts on a continental scale, the Joint Research Center (JRC) of the European Commission (EC) is developing a prototype of the European Drought Observatory (EDO). A multidisciplinary set of indicators has been introduced, which is used within EDO to continuously monitor the various components of the environment potentially affected by this hazard (soil, vegetation, etc.) in order to obtain a comprehensive and up-to-date picture of the situation. Two indicators produced under EDO compare yield statistics to assess the effects of drought events on agricultural production. The test area is Spain, which is often suffered from a severe and prolonged drought. The results show that yields are significantly reduced in line with the drought events found by the indicators.

As drought is a slow-growing phenomenon affecting the whole water cycle (e.g. soil moisture, river basins, water levels in lakes, reservoirs, and groundwater) and has direct effects on vegetation, all these components must be monitored continuously throughout a long period of time. Other important aspects of adequate drought management are drought risk analysis (i.e. the likelihood of drought occurring to a certain degree, severity, and duration) and social vulnerability to drought. They are the basis of a detailed risk assessment. Finally, in the short and medium-term, forecasting the occurrence and likely development of droughts, as well as forecasting and analyzing the likely effects of climate change on drought hazards in different regions, are important to support the development of effective management plans on land.

At the beginning of the last century, attempts began to define the concept of drought, especially agrometeorological drought, and ways to identify it. Based on the common for all types of drought - precipitation deficit, almost all indices are related to them - a different number of days without precipitation, and later they refer to the consumption of water from soil and vegetation - evapotranspiration. In 1965, Palmer published its model for determining three types of drought - meteorological (PDSI), hydrological (PDHI), and agricultural (Z-index). A historical review and analysis of the indices used in the United States for different types of drought by [1] shows that the Crop Moisture Index (CMI), the Palmer Moisture Anomaly Index, is the most widely used to study the agricultural type of drought.

In relation to significant a problem with identifying drought – duration and intensity in 2013, World Meteorological Organization (WMO) and the Global Water Partnership (GWP) started an Integrated Drought Management Programme (IDMP), Integrated Drought Management Programme. The first step of drought management is Monitoring and Early Warning Systems. The most commonly used drought indicators/indices for detecting the drought are shown in the Handbook of Drought Indicators and Indices [2]. Experience has shown that it is difficult to establish common indicators. Due to the complexity and variability of drought depending on climatic and geographical conditions, it is appropriate to work on different indicators to be included in drought monitoring and warning systems. Selecting of the indicators has to consider timely detection of drought to realize appropriate communication with users and coordination to mitigate of effect. The characteristics of climate in space and time also have to have in mind to determine drought onset and termination. One of the most important things in the selection of indicators is the availability of information.

Every year, huge financial resources are invested in agriculture to maintain a plant-friendly water and nutrition regime. Many of them are lost due to the low level or lack of scientific management of activities in every production field in our country and in most countries of the world. The losses are due to inefficient use of energy, water, fuel, and human labor, as well as data from the meteorological network. As a result, agricultural fields are an intensive source of pollutants that damage ecosystems.

2. Drought monitoring in Bulgaria

The studies on the frequency of dry periods with different duration in Bulgaria have a special interest in view of the dry character of climatic conditions in the agricultural regions. An in-depth study of the conditions for the occurrence of drought and drought periods in Bulgaria during 1983–1994 was made by Kincaid and Heermann [3]. There is defined that the driest time during the year occurs most frequently in the period from August to October, and the areas with the highest number of droughts are the Seaside and Southern parts of the lowland of Tundja, Maritsa, and Struma [3, 4]. A similar trend is observed in the last 20 years - increased frequency and the dry period compared to the modern climate, especially in the Thracian Lowland and Northeastern Bulgaria [5, 6]. The most used agroclimatic indices in historical scale are Selianinov's HTK, De Martone Index, Thornthwaite Index, the balance of atmospheric humidity, aridity index, and so on. They are used for a comprehensive assessment of temperature and humidity conditions in the area under consideration. The most commonly used is aridity index because it gives an idea of the real water deficit for a certain period. Dilkov [7] found that in the period of spring, wheat growth evaporability-precipitation balance values are exceeding

−200 mm in each 2 to 4 years of 10 years. More recent studies [8, 9] that the evaporability-precipitation balance in the spring growing season for the period 1971–2000, the range between −223 mm and + 15 mm, the largest deficit of water resources is observed in some areas of the Thracian lowland and especially agricultural lands around Svilengrad, Ivailo, Plovdiv (−180 mm). The values of De Martone in agricultural areas ranged from 20 to 40 - mm/°C, [10–12] which defines the terms as moderately moist, HTK Selyaninov of about 1, and wheat yields are obtained when values of the De Martone than 30 mm/°C. Comprehensive assessment of conditions in the areas of agricultural production shows that in the region of Thracian lowland and Dobroudja to obtain high yields of wheat is necessary to compensate for the water deficit, is to conduct additional irrigation at critical growth and yield formation of cultivations. The change in the deficit in the root zone of the specific soil sown with a certain crop during the current growing season differs significantly from that calculated by climatic methods, which are based on average multi-year data. Estimates of the impact of climate change during the period 1961–2000–2010 on the soil moisture in wheat cultivation on six soil types were made and results were obtained for 24 agricultural stations in the country [13].

To characterize the hydrothermal conditions for growing autumn and spring crops in the country, the Selyaninov Hydrothermal Coefficient (SHC), De Matron (AI_{DM}), and the potential evapotranspiration on Thornthwaite (ETP) were used by processing data from 42 climatic and agrometeorological stations [10–12].

The influence of climatic changes on the atmospheric humidity, the evaporation from a free water surface, and the reference evapotranspiration on the territory of Bulgaria have been studied. The most unfavorable changes in the evaporating conditions have occurred in the Petrich-Sandanski climate region, as well as in the Central and Eastern part of the Danube Plain [10–12]. 30-year FAO reference evapotranspiration rates were obtained using the Penman-Monteith (ETo) equation for 30 agrometeorological stations. A summary is made by climatic regions [14].

The single and combined effect of the meteorological parameters of the FAO Penman-Monteith equation on the estimates for the reference evapotranspiration during different sub-periods of the potential vegetation period (PVP) related to the physiological development of crops and their sensitivity to moisture is analyzed. One-factor and two-factor correlation analysis was conducted for 30 agrometeorological stations on the territory of agricultural production in Bulgaria [15].

Maps for the average annual spatial distribution of the atmospheric moisture balance (BAO = sum of precipitation minus reference evapotranspiration) are presented, as well as for the 30-year change of this deficit, and the areas with the most unfavorable conditions for agricultural crops are indicated. The need to update the irrigation regime of crops is justified [16, 17].

Maize in Bulgaria is a crop that suffers from water deficiency during its most sensitive phases in terms of water - sweeping-squeezing and squeezing-milk maturity. Data from the period 1971–2000 from 9 representative agrometeorological stations were processed. The climatic moisture supply of corn for grain, grown on four types of chernozems in Northern Bulgaria - typical, leached, carbonate, and degraded, is analyzed [8].

Comparative study of drought use monitoring of 3 indices is made in National Institute of meteorology and hydrology (NIMH). For detection of meteorological drought during 2009 WMO recommended the Standardized Precipitation Index (SPI) as a main meteorological index for observing and analyzing the drought conditions [18]. SPI *transform the precipitation totals for a certain period in a standard normal distribution.* In the intensity scale of SPI, the positive and negative values correlate with the up normal rainfall and drought. McKee et al. [19] determined that the drought started at values of SPI ≤ -1, but in many cases are used different values between 0 and -1 or below -1 [20]. In many countries, including and these in Southern and South-Eastern Europe, classifications are used, including the category "normal" for SPI values in the range of ± 0.5.

The Soil Drought Index (SMI) developed by the HPRCC (High Plains Regional Climate Center) determines the intensity of drought by assessing the available water available to plants in the soil [21] relative to its maximum quantity for a given soil type. To calculate it, the measured soil moisture in agricultural crops is used, which allows to determine the degree of drought for a particular crop. It characterizes soil drought from normal to extreme, with the degree of drought increasing with decreasing index.

The monitoring is made with monthly data which means that the information has a diagnostic character.

Fundamental research is needed to gain knowledge about the root causes of the processes taking place in agroecosystems.

Inefficient management of activities in each field stems from the inability to easily and cheaply obtain current data on water supply (or deficit) in the soil by applying local (point) methods for measuring soil moisture and the climatic and remote methods used requiring highly qualified specialists [22]. This difficulty is related to the huge number of agricultural fields. It was found that when measuring 2 places in 10 depths for an area of 50.0 ha, the error was 79.6%. In order to reduce the error to 7.0%, it is necessary to make 50 measurements of the soil moisture profile, including 10 points in depth [23]. These results show that the representativeness of the experimental data obtained using these methods is a serious problem. Their application is limited within experimental stations, as well as for instrument calibration, regardless of the developed methodologies to reduce the number of measurements. This is confirmed by [3], who showed that for each field it is necessary to obtain representative data on water deficit over three days for the entire growing season. For example, for only one cornfield with an area of 50 ha, we need on average about 25,000 point measurements for a vegetation period lasting 150 days.

3. Material and methods

Solving various global tasks, such as the improvement of irrigation systems is related to determining the agroclimatic resources for growing cereals under irrigated and non-irrigated conditions requires the processing of data from meteorological and agrometeorological observations in all parts of the country. Observations from two or three climate and agroclimatic stations were taken in some larger regional centers. Thus, the territory of the country, which consists of 28 districts, was covered with 70 stations from the meteorological and agrometeorological networks of NIMH. The location of the stations is chosen so that the distance between them is approximately 20–30 km. Such a distance between the measuring points ensures the correctness and representativeness of the field of the measured elements related to the temperature conditions and their derivative characteristics. These include the maximum, minimum, and average daily temperatures and the 24-hour amount of precipitation, and the derived characteristics include active and effective temperatures reported above a certain biological threshold. Most hydrothermal indices reflecting the conditions of moisture and the characteristics of evaporation from the soil and crops were included here, i.e. evapotranspiration.

Under these conditions, there are suspicions about the interpretation of the precipitation field, but its structure is not the subject of this study. The latter is measured with sufficient accuracy, where the error due to the slightly greater distance between the measurement points is largely compensated by the number of measurements of water reserves in soil layers to a depth of 1 m.

In addition to the temperature and precipitation to characterize the conditions of drought and drought in any part of the country, the working database included data on the average daily values of agility of water vapor, the relative humidity of 2 m in the weather cell, wind speed and the duration of sunshine.

The working database is created with the data for the average daily values of maximum (Tmax), minimum (Tmin), average daily (Tav) temperatures, the agility of water vapor (E), relative humidity (F), wind speed (w), the duration of sunshine (s), and the sum of precipitation (r) for all 70 stations, **Figure 1**.

With the mentioned data, Excel spreadsheets were formed, as a separate file was created for each station, and the data cover a 30-year measurement period - from 1986 to 2015 inclusive. In this form, the control of the data, the marking of missing data, and their recovery were carried out. Statistical data processing was also performed - mean of each of the series, standard deviation, variance, standard deviation, median, mode, and type of data distribution.

In addition, a single database was created with the Excell® files. In the middle of the database are performed all calculations and selective data processing - the climatic value of meteorological elements, calculation of coefficients and indices reflecting the hydrothermal conditions and their spatial distribution throughout the country.

The methodology of the Joint Research Center in Ispra, Italy, and the General Directorate of Agriculture of the European Commission was used to characterize the drought conditions. According to her, the Aridity Index (AI) has recently been widely

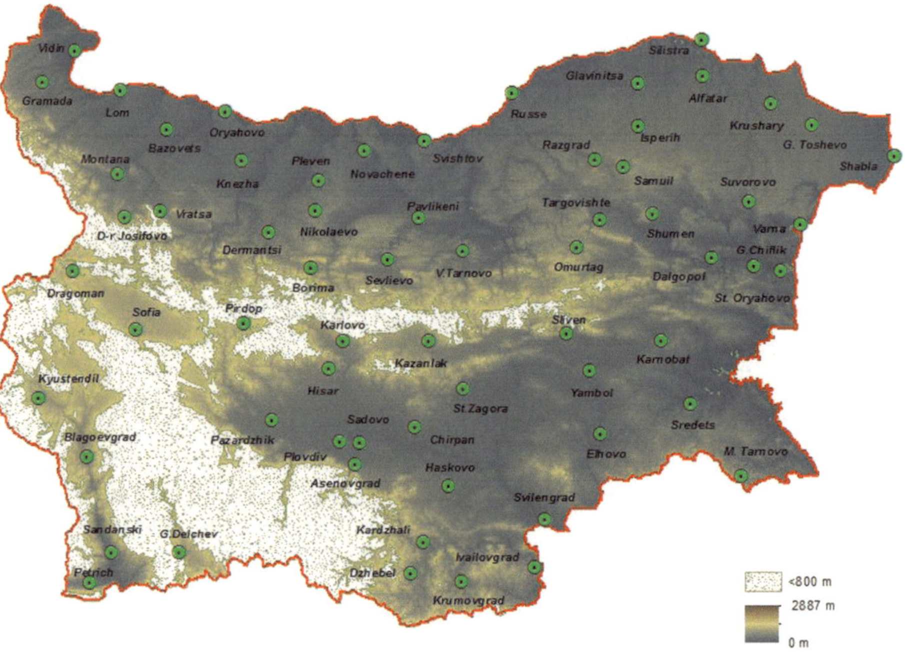

Figure 1.
Spatial distribution of meteorological stations on the territory of the country.

used to assess the conditions of drought and drought in agriculture. The calculation of the dryness index is applied by formula (1), recommended by [24], in the Methodology for identifying areas with natural constraints, described in the JRC Technical by calculating the dryness index:

$$\mathrm{AI}_{\mathrm{UNEP}} = \sum \mathrm{r}/\mathrm{ETP}, \tag{1}$$

where AI - drought index, Σr - annual amount of precipitation; and ETP - the sum of the annual potential evapotranspiration. All values of the drought index.

Potential evapotranspiration - ETR, as defined by FAO-56, is the evapotranspiration from a grass surface with a standard height of 8–15 cm of plants that are actively growing, completely shading the soil surface and not experiencing water shortages [25, 26]. In recent years, a new calibrated method for calculating potential evapotranspiration has emerged worldwide - the FAO Penman-Monteith method [27, 28].

Potential evapotranspiration (ETP) by this method is determined by the equation:

$$ET_o = \frac{0,408\Delta(R_n - G) + \frac{\gamma 900 U_2 (e_a - e_d)}{T+273}}{\Delta + \gamma(1 + 0,34U_2)} \tag{2}$$

where.

ETP potential evapotranspiration [mm d^{-1}],

R_n и G net radiation and heat flow in the soil [MJ m^{-2} d^{-1}],

T the average daily air temperature в [°C],

Δ water vapor pressure gradient [kPa °C^{-1}],

γ psychrometric constant [kPa °C^{-1}],

e_a *u* e_d deficit of air saturation with water vapor at standard altitude (at the level at which the meteorological measurements are performed) - 2 m [kPa],

U_2 wind speed at a reference height of 2 m [m s^{-1}].

Potential evapotranspiration (ETP) was used to assess the saturation of the atmosphere with water vapor, and real evapotranspiration (ETR) was used to assess the behavior of the plants concerned under certain evaporative conditions in the atmosphere.

4. Results and discussion

The annual course of potential evapotranspiration in certain meteorological conditions is determined by the biological characteristics and physiological development of crops, their water needs at each stage of their development. Physical conditions have the effect of increasing or decreasing evapotranspiration, while physiological conditions associated with aging have the effect of limiting the influence of external factors and regulating the evaporation process.

The study used daily data, which were mentioned above for the period 1961–2015. The study performed the following:

- With the given data calculations have been made with the Penman-Monteith Eq. (2) FAO56 and the CropWat® model. Daily values of the potential evapotranspiration for each station for the 30-year period were obtained;

- The amounts of ETR for each station are calculated for three important periods for agricultural production and irrigation in our country: 1) the period from sowing to emergence vegetative development and overwintering of autumn crops from October of the previous to March of the following year; 2) the period of ripening of winter cereals April-June and 3) the period of irrigation of spring crops – July-August;

- The trends of the potential evapotranspiration for the research period by representative stations and administrative regions have been obtained. The results for both the potential vegetation and the traditional period of irrigation in our country – June-August are presented.

The differences between the final and initial values of the trends representing the change in the potential evapotranspiration over the 30-year period have been calculated.

The obtained results for the potential evapotranspiration by years were averaged for the study period, the minimum and maximum values, standard deviation, coefficient of variation, steepness, and median for each point of the agricultural territory of the country were determined, **Table 1**.

Station/Statistics	Long-term Average	St. deviation	Min. value	Max. value	Var. coefficient	Kurtosis	Median
Knezha	935.87	72.35	793.17	1171.97	59.85	0.60	921.22
Pleven	979.75	78.62	841.43	1225.10	59.37	0.91	982.17
Pavlikeni	976.44	155.83	773.34	1321.79	135.88	−0.87	923.22
Russe	1102.42	112.36	928.83	1353.54	92.28	−0.54	1113.00
Targovishte	925.91	91.62	735.78	1147.28	73.89	−0.21	930.00
Ispeih	912.57	73.15	795.81	1073.52	61.16	−0.75	907.60
Shumen	955.58	63.76	832.77	1146.72	51.55	0.11	951.28
G.Toshevo	923.00	79.69	805.44	1112.34	64.77	−0.09	913.33
Montana	963.01	132.49	760.36	1278.14	117.33	−0.88	910.44
Vidin	895.78	57.81	675.95	1013.36	40.84	4.33	901.75
Average North Bulgaria	957.03	91.77			75.69	0.26	945.40
Karnobat	945.62	53.63	1186.18	1186.18	42.96	0.04	944.92
Elhovo	977.44	63.15	1084.43	1084.43	47.79	1.54	980.96
Sliven	1029.60	71.63	1253.60	1253.60	56.96	0.03	1026.75
Chirpan	990.31	60.72	1301.45	1301.45	48.51	0.84	988.93
Kazanlak	932.96	63.66	1005.92	1005.92	51.36	−0.30	934.53
Plovdiv	1016.31	76.27	997.63	997.63	60.15	0.45	1014.55
Sandanski	1115.16	78.78	1074.23	1074.23	64.49	−0.45	1109.64
Kyustendil	903.10	53.10	1093.01	1093.01	43.23	−0.82	905.92
Sofia	894.76	48.30	1168.99	1168.99	40.25	−0.56	893.63
M.Tarnovo	914.19	61.95	1213.74	1213.74	48.63	0.02	908.51
Average South Bulgaria	971.94	63.12			50.43	0.08	970.83

Table 1.
Statistical characteristics for 20 representative stations in northern and southern Bulgaria.

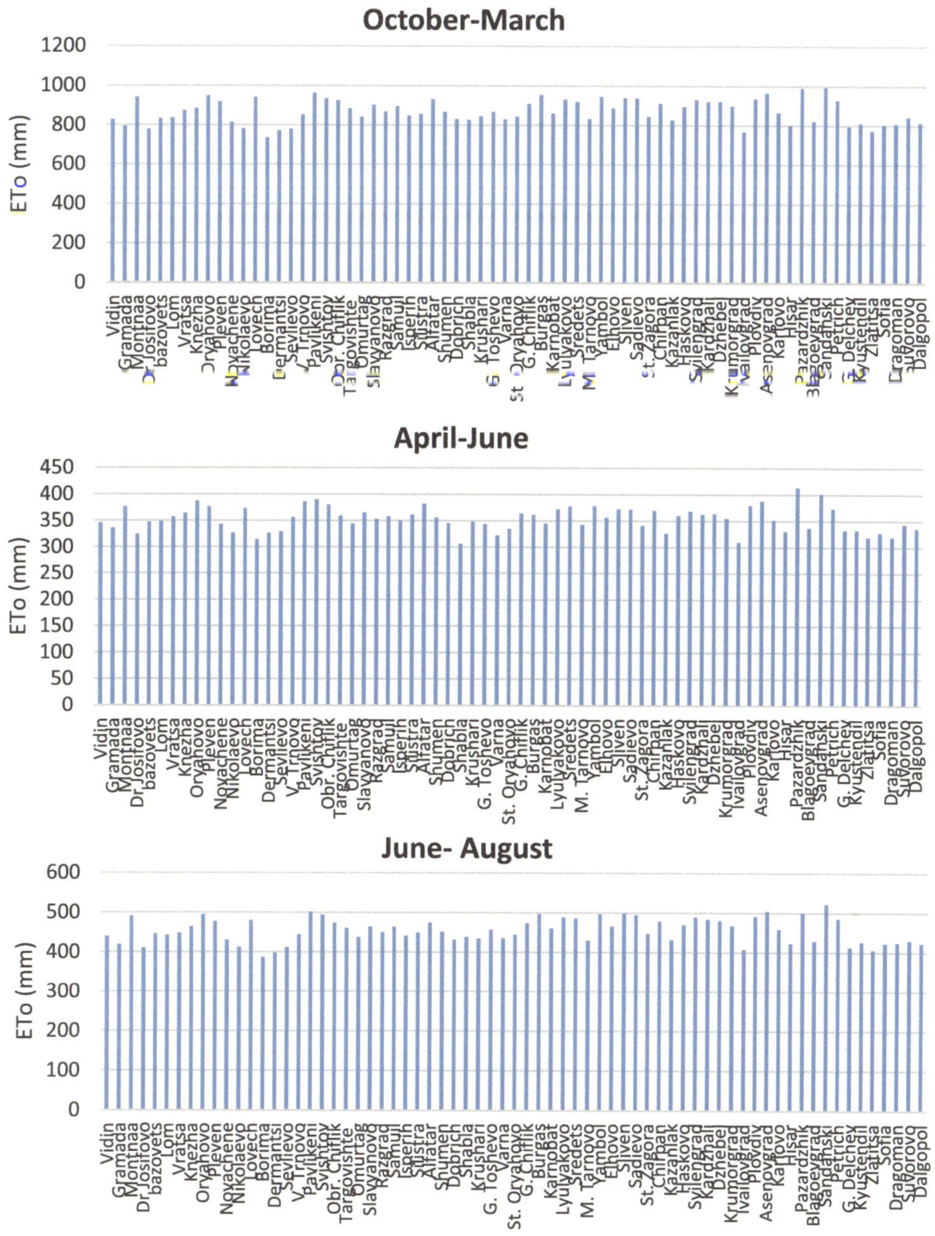

Figure 2.
Average long-term values of the potential evapotranspiration by stations and periods of crop development for the period 1961–2015.

The graphic materials and simulations cover 18 representative stations to describe in the most plausible way the conditions in the six regions into which we have divided the country.

The values of the potential evapotranspiration by stations and periods of crop development were also calculated, as already noted. During the period October–March the predominant value of ETP is above 800 mm–850 mm, higher than 900 mm is the potential evapotranspiration in Gramada, Nikolaevo, Pavlikeni, Hisar,

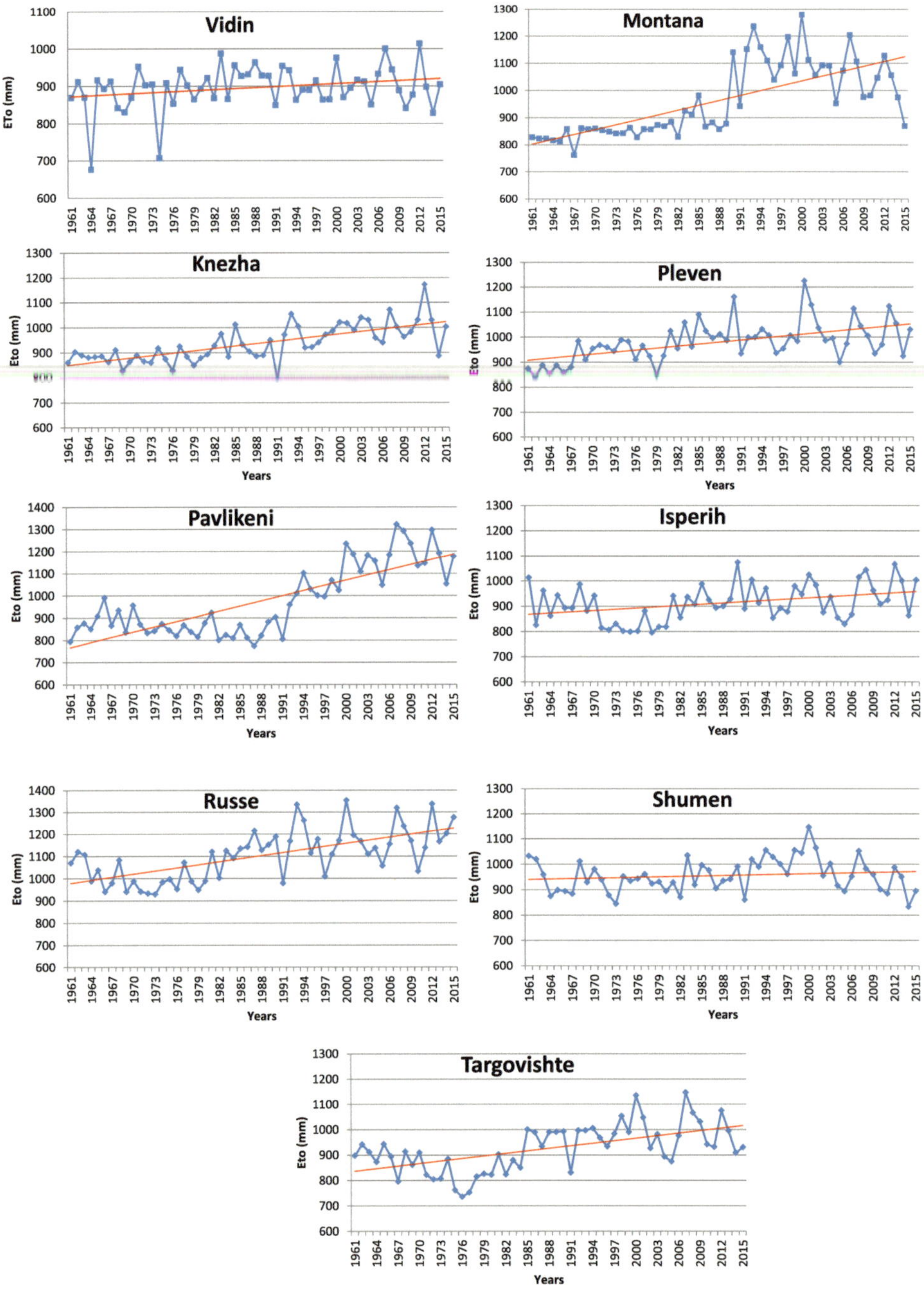

Figure 3.
Trends of potential evapotranspiration in northern Bulgaria.

Sandanski, Svishtov, and Burgas. The lowest values of ETP were obtained in Lovetch, Borima, Dermantsi, Sevlievo, and Ivaylovgrad, a less than 800 mm, **Figure 2**.

The values of the potential evapotranspiration by stations and periods of crop development were also calculated, as already noted. During the period October–March, the predominant value of ETP is close to 800 mm.

The dynamics and the trend of change of the potential evapotranspiration in Northern and Southern Bulgaria during the investigation period are presented in **Figures 3** and **4**. It can be seen that in almost all representative stations there is a tendency to increase the potential evapotranspiration. While in the northern regions this process is clearly visible, in the southern regions there is diversity, and in the northwestern and northeastern parts of the country, the increase in ETP is well expressed, in the central regions this process oscillates around an average and only in Kanzanlak, Kyustendil, and Sofia to a slight reduction in ETP. Precise analysis shows that this is due to an increase in the amount of precipitation. These results are also confirmed by the data of the conducted significance test by the Mann-Kendall method, **Table 2**.

The values of the long-term average monthly potential evapotranspiration for 18 representative stations from Northern and Southern Bulgaria, which correspond to the transitional-Continental and transitional-Mediterranean type of climate are presented in tabular and graphical form in **Table 2** and in **Figure 5**.

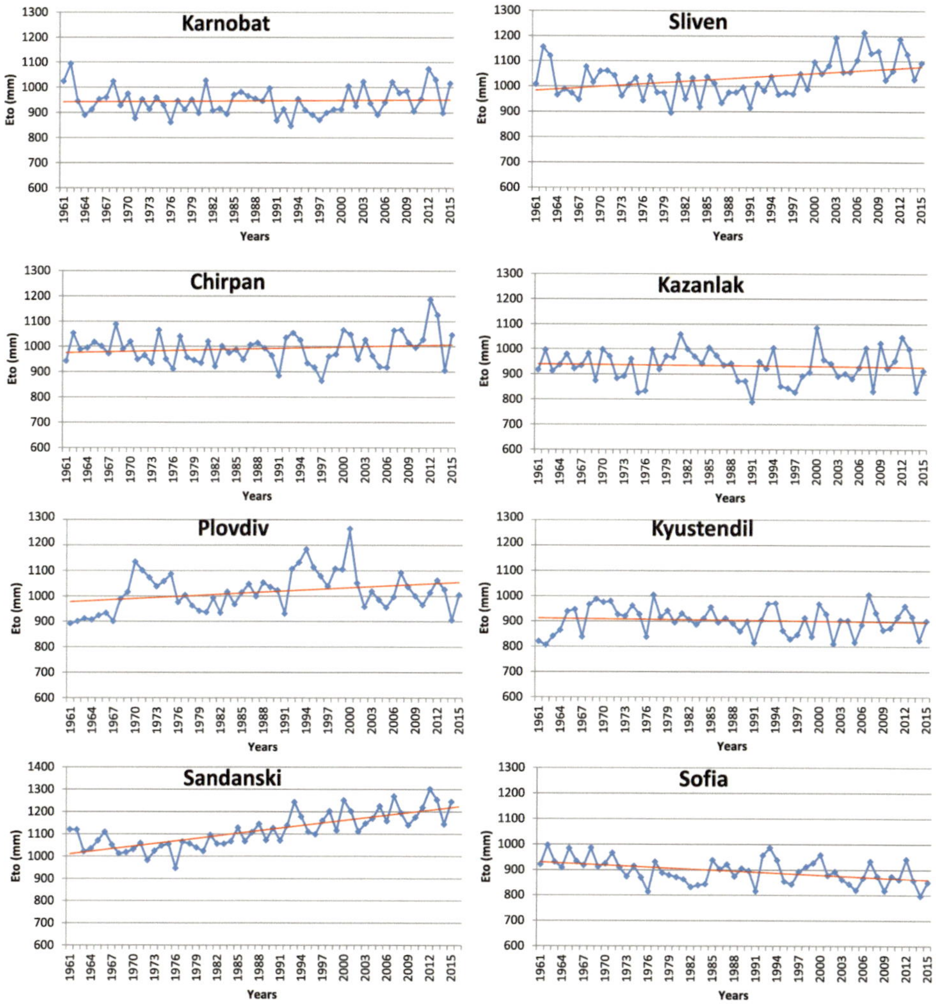

Figure 4.
Trends of potential evapotranspiration in southern Bulgaria.

III-X	Z criteria	Sign	IV-VI	Z criteria	Sign	VI-VIII	Z criteria	Sign
Vidin	1.2	+	Vidin	0.5		Vidin	2.6	**
Montana	6.4	***	Montana	6.5	***	Montana	6.4	***
Knezha	5.2	***	Knezha	3.5	***	Knezha	4.0	***
Pleven	3.7	***	Pleven	2.9	**	Pleven	3.1	**
Novachene	−1.1	+	Novachene	−1.1		Novachene	−0.3	
Nikolaevo	−0.3	+	Nikolaevo	−0.3		Nikolaevo	0.6	
Pavlikeni	5.5	***	Pavlikeni	4.6	***	Pavlikeni	5.5	***
Obr. Chiflik	3.9	***	Obr. Chiflik	3.8	***	Obr. Chiflik	3.1	**
Targovishte	4.1	***	Targovishte	3.2	**	Targovishte	3.8	***
Isperih	2.4	*	Isperih	2.3	*	Isperih	3.2	**
Shumen	1.2	+	Shumen	1.6		Shumen	1.4	
Krushary	2.0	*	Krushary	2.0	*	Krushary	1.0	
G. Toshevo	3.9	***	G. Toshevo	2.6	**	G. Toshevo	3.5	***
Karnobat	0.68	+	Karnobat	1.8	+	Karnobat	2.0	*
M Tarnovo	1.80	+	M Tarnovo	1.4		M Tarnovo	2.0	*
Elhovo	−0.63	+	Elhovo	0.3		Elhovo	0.1	+
Sliven	2.84	**	Sliven	2.5	*	Sliven	2.5	*
Chirpan	0.38	+	Chirpan	0.8		Chirpan	1.2	+
Kazanlak	−0.66	+	Kazanlak	−0.3		Kazanlak	0.1	+
Plovdiv	2.12	*	Plovdiv	1.4		Plovdiv	2.6	**
Blagoevgrad	−2.85	**	Blagoevgrad	−1.7	+	Blagoevgrad	−1.6	
Sandanski	5.74	***	Sandanski	4.0	***	Sandanski	5.1	***
Petrich	−2.42	*	Petrich	−2.3	*	Petrich	−2.1	*
Kyustendil	−0.94	+	Kyustendil	−0.7		Kyustendil	0.9	+
Sofia	−2.48	*	Sofia	−0.2		Sofia	−0.2	

*Levels of significance *** - α = 0.001; ** - α = 0.01; * - α = 0.05 and + — α = 0.1.*

Table 2.
Levels of significance of the trends determined by the Mann-Kendall test for change of ETP by agricultural crops and periods of development.

The average multi-year monthly values of the potential evapotranspiration were calculated, which are shown in **Table 3**. The highest monthly values of ETP were reported in the stations Sandanski-198 mm and Sliven-182 mm. These values were reported in July, the warmest month of the year, and the lowest values of this indicator for the same period of the year are in the Krushari – 148 and Sofia-152 mm.

The average long-term values of soil water consumption through evapotranspiration in the period 1961–2015 in Northern Bulgaria are increasing, compared to the reference period. This increase is most noticeable in the northeastern region during the autumn-winter period October–March - 117 mm,

DOI: http://dx.doi.org/10.5772/intechopen.102391

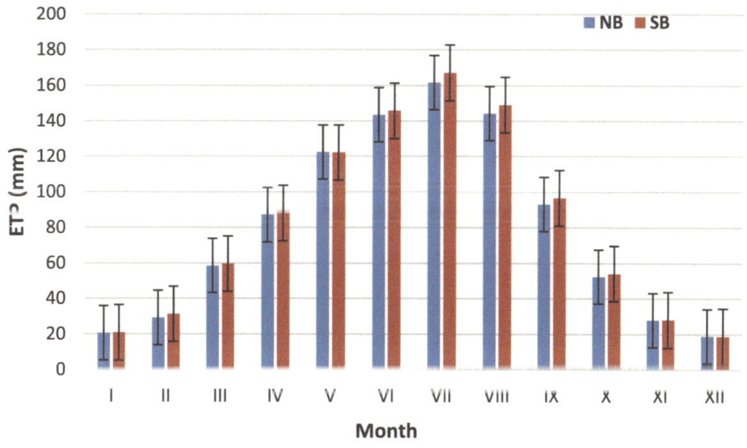

Figure 5.
Average multiannual values of ETP (mm) for northern and southern Bulgaria, which correspond to the transitional-continental and transitional Mediterranean types of climates and for 1986–2015.

Months/Stations	I	II	III	IV	V	VI	VII	VIII	IX	X	XI	XII
Vidin	18	26	55	83	121	143	160	138	86	45	21	15
Montana	24	34	67	93	129	156	175	160	103	57	31	22
Knezha	19	28	60	90	126	150	166	149	95	51	27	17
Pleven	21	30	63	95	130	154	171	154	98	55	28	19
Novachene	14	23	52	84	120	142	154	138	84	45	20	12
Nikolaevo	15	25	52	81	113	134	148	132	81	44	21	13
Pavlikeni	22	33	63	97	132	157	178	164	106	61	32	20
Obr. Chiflik	20	29	63	97	134	150	171	153	102	55	28	19
Targovishte	21	30	61	90	124	146	168	147	96	53	28	19
Isperih	20	26	56	87	124	141	159	142	92	50	27	18
Shumen	22	31	59	89	124	144	163	145	93	53	29	21
Krushary	18	26	55	88	123	137	155	141	92	51	27	17
G. Toshevo	20	27	54	85	120	139	165	152	97	55	28	20
Karnobat	19	28	54	83	118	144	167	150	93	49	24	17
M Tarnovo	23	30	54	86	119	138	155	138	90	53	31	23
Elhovo	19	31	58	89	124	145	167	153	97	55	27	18
Sliven	23	33	60	90	128	156	182	162	105	58	29	22
Chirpan	17	29	59	90	126	152	171	157	103	52	24	14
Kazanlak	21	30	56	81	113	134	157	142	94	53	27	18
Plovdiv	23	34	63	94	128	157	177	158	103	56	30	20
Blagoevgrad	17	29	57	86	115	137	157	136	86	49	25	16
Sandanski	25	40	75	103	140	174	198	174	115	63	31	22
Petrich	20	32	64	92	130	154	177	158	103	58	26	17

Months/Stations	I	II	III	IV	V	VI	VII	VIII	IX	X	XI	XII
Kyustendil	17	28	58	85	114	135	156	137	84	45	22	15
Sofia	17	27	55	82	112	134	152	137	86	46	23	15

Table 3.
Average long-term values of potential evapotranspiration (ETP) for some representative agrometeorological stations by months for the period 1961–2015.

after it is the northwest - 114 mm, followed by the north-central region - 102 mm. An increase in evapotranspiration is also present in the other two periods, as in the first of them the values of the increase are 27 mm–44 mm, and in the period June–August these values are 48 mm–68 mm, as the higher values refer to the northwest, and the lower ones for the northeastern region. On average for Northern Bulgaria, the increase in evapotranspiration for the entire growing season is 201 mm, **Table 3**.

Soil water consumption by evapotranspiration in Southern Bulgaria for the study period is similar and a bit lower to this in Northern Bulgaria, **Table 3**. Summarizing the results of the comparative study we should note an overall increase in potential evapotranspiration throughout the country compared to the reference period 1961–1990. The most noticeable is the decrease of ETP during the autumn-winter period in the high valley fields of Western Bulgaria −61 mm; an Increase of ETP is observed in the southwestern region - 108 mm. During the period April–June the decrease is observed in the high valley fields of Western Bulgaria −22 mm, followed by the south-central region - 36 mm. An increase in ETP is also observed in the period June–August, as again the highest values belong to the high valley fields in Western Bulgaria with 122 mm and the southwestern region - 59 mm. The average values of increase in soil water consumption through evapotranspiration in Southern Bulgaria in the three periods is 117 mm in the autumn-winter period; 36 mm in spring and summer and 62 mm in summer.

Obtained values of the AI index are analyzed in time and space, as a result the dry and wet years during the studied period are determined (criterion - the number of stations with AI≤0.6 for the agricultural zone of the country is more than half), and also regions in which AI≤0.6 by years (criteria - the number of years with AI ≤0.6 to be greater than or equal to 7, which is in accordance with the cited above methodology. By applying the criteria AI≤0.6 dry year; 0.6 ≤ AI≤1.0 normal year and AI≥1.0 wet year we are defined the last 30 years as dry, wet, and normal years according to the values of the drought index as follows:

Dry years are - 1985, 1986, 1988, 1989, 1990, 1992, 1993, 1994, 2000, 2001, 2006, 2008, 2013, and 2015.

Wet years are - 2002 and 2005, 2012, and 2014.

Normal years are - 1987, 1989, 1991, 1992, 1995, 1996, 1997, 1998, 2003, 2004, 2007, 2009, 2010, and 2011.

As a result of the simulations, the average multi-year dates of the beginning of the depletion of water reserves below the lower limit of optimal moisture (70% of AWC) were obtained, which requires the first irrigation of corn crops. For Northern Bulgaria, the deadlines are from May 31 to June 17, and the increase of the date is from west to east, **Table 4**.

In Southern Bulgaria, the average dates of depletion of water reserves occur one week earlier - May 24, and in the sub-Balkan fields and high valley fields of Western Bulgaria, this happens in the period June 20–26, **Table 5**.

Production areas	October–March 1961–1990/1986–2015/trend	April–June 1961–1990/1986–2015/trend	June–August 1961–1990/1986–2015/trend
North Bulgaria			
Northwest	736/850/+114	306/351/+44	378/445/+68
North Central	750/852/+102	320/353/+32	394/446/+52
Northeast	695/812/+117	310/337/+27	383/431/+48
Average for Northern Bulgaria	727/838/+111	312/347/+34	385/441/+56
South Bulgaria			
Southwest	805/913/+108	329/358/+29	419/478/+59
South Central	790/875/+86	324/359/+36	396/443/+47
Southeast	791/878/+88	325/354/+29	415/464/+49
High fields of Western Bulgaria	908/847/−61	369/347/−22	388/367/−22
Sub-Balkan fields	791/847/+56	317/350/+33	411/445/+35
Average for Southern Bulgaria	775/892/+51	323/358/+21	401/464/+34

Table 4.
Sums of potential evapotranspiration (ETP) (mm) for character periods by agroindustrial regions of the country.

Station	Date of sowing	Date of first watering
Vidin	20.04	31 V
Montana	27.04	11 VI
Knezha	25.04	11 VI
Pavlikeni	25.04	14 VI
Razgrad	29.04	17 VI
Shumen	29.04	17 VI
Karnobat	26.04	12 VI
M. Tyrnovo	20.04	7 VI
Sliven	20.04	30 V
Kazanlyk	30.04	20 VI
Chirpan	15.04	26 V
Plovdiv	15.04	24 V
Sofiia	28.04	26 VI
Sandanski	15.04	25 V

Table 5.
Average long-term dates of depletion of water in the soil below 70% of AWC and determination of the need for irrigation in maize-grain crops, FAO group 400–500.

5. Conclusions

This study presents part of the results related to the need for continuous monitoring of potential evapotranspiration (ETP) and monitoring of trends in its change. Within this study, the parameters of ETP for the period 1961–2015 were obtained for 72 stations from the agricultural territory of Bulgaria, and also the results for some stations representative for the agricultural production are shown. The more important conclusions are the following:

1. The values of average mean temperatures, amounts of precipitation, and amounts of potential evapotranspiration representative for agricultural production during the study period 1986–2015 have a positive trend compared to the same values during the reference period 1961–1990;

2. The data obtained for the period 1986–2015 show positive values of deviations and an increase in evapotranspiration throughout the country with a tendency for this process to continue to increase;

3. The uneven nature of the distribution of precipitation by seasons is intensifying and the dry winters, dry beginning of spring, rainy June, and prolonged 70–90 day summer drought, which more and more often turns into autumn drought, become more frequent. These features of humidification conditions are a serious challenge for selection specialists in creating varieties that can withstand periods of drought and drought;

4. The analysis of the results obtained at this stage is reason to recommend in the coming years a gradual but large-scale increase in investment for the construction of modern irrigation systems. The results of the climate scenarios for the next 20–30 years show that the trend of the observed changes will continue, which will greatly hinder and reduce the efficiency of individual branches of agricultural production under natural conditions of humidification;

5. Need for solid support and rapid development of precision and organic farming through digitalization, but with the leading participation of experts in agrometeorology, climate change, and agronomy.

Acknowledgements

This work was supported by the Bulgarian Ministry of Education and Science under the National Research Programme "Healthy Foods for a Strong Bio-Economy and Quality of Life" approved by DCM # 577/17.08.2018.

Author details

Valentin Kazandjiev*, Veska Georgieva, Petia Malasheva and Dragomir Atanassov
National Institute of Meteorology and Hydrology, Sofia, Bulgaria

*Address all correspondence to: valentin.kazandjiev@meteo.bg

References

[1] Keyantash J, Dracup JA. The Quantification of Drought: An Evaluation of Drought Indices. Bulletin of the American Meteorological Society. 2002;**83**:1167-1180. DOI: 10.1175/1520-0477(2002)083<1191:TQODAE>2.3.CO;2

[2] Available from: https://www.droughtmanagement.info/literature/GWP_Handbook_of_Drought_Indicators_and_Indices_2016.pdf

[3] Kincaid DC, Heermann DF. Scheduling Irrigation Using a Programmable Calculation. Agricultural Research Service, USDA; 1973

[4] Knight G, Raev I, Staneva M. Drought in Bulgaria. Vol. 284. London: BAS, Routledge; 2003

[5] Alexandrov V, Radeva S, Koleva E. Utilization of SPI, PDSI, and RDI as drought indicators in South Bulgaria. 11th International Multidisciplinary Geoconference Proceedings Volume II; Albena, Bulgaria. 2011. pp. 969-976

[6] Booher LJ. Surface Irrigation. Rome: FAO; 1974

[7] Dilkov D. Moisture consumption and moisture supply of wheat in our country. Tr. IHM. 1960;**VIII**:3-48

[8] Kazandjiev V, Moteva M, Georgieva V. Climate Change, Agroclimatic Resources and Agroclimatic Zoning of agriculture in Bulgaria. Agricultural Engineering. 2010;**3**:109-116

[9] Moteva M, Kazandjiev V, Georgieva V. Assessment of the contemporary hydro-thermal conditions for crop growing in Bulgaria. Proceedings on the International Conference on “Soil Tillage and Ecology”, 1–5 September; Albena, Bulgaria. 2009. pp. 134-141

[10] Moteva M, Kazandjiev V, Georgieva V. Climate change and the hydrothermal and evapotranspiration conditions in the panning regions in Bulgaria. Proceedings of 14-th International water Technology Conference, IWTC Cairo - Egypt; 2010a 12 p

[11] Moteva M, Kazandjiev V, Georgieva V. Impact of Climate Change on Soil Water Table in Sofia Region, Bulgaria. Proceedings of the International Conference on Water Observation and Information System for Decision Support, Ohrid, Macedonia, CD version: 25–29 May; 2010b

[12] Moteva M, Kazandjiev V, Georgieva V. Influence of climate change on the reference evapotranspiration in Bulgaria. Plant Science. 2010c;**2**:181-186

[13] Georgieva V, Moteva M, Kazandjiev V. Impact of Climate Change on Water Supply of Winter Wheat in Bulgaria. Agriculturae Conspectus Scientificus. 2007;**72**(1):39-44

[14] Moteva M, Kazandjiev V, Georgieva V. Study of the reference evapotranspiration according to FAO Penman-Monteith on the territory of Bulgaria. Agricultural Machinery. 2008;**5**:26-33

[15] Kazandjiev V, Moteva M, Georgieva V. Interaction between the Meteorological Parameters in FAO 56 Penman-Monteith Equation and the Reference Evapotranspiration Estimates. Conference “Global Environmental Change–Challenges to Science and Society in Southwestern Europe” CD version; 2008

[16] Kazandjiev V, Moteva M, Georgieva V, Dimitrov P. Atmospheric water supply for the potential evapotranspiration in Bulgaria. Lviv, Ukraine, CD Version: ICID 23rd ERC 2009 "Progress in Managing Water for Food and Rural Development", 18–24 May; 2009

[17] Kazandjiev VS, Georgieva VA. In: Meena RS, editor. Chapter 4. Changes in the Agro-Climatic Conditions in Bulgaria at the End of the 20th and the Beginning of the 21st Century book Agrometeorology. BHU, India: Department of Agronomy Institute of Agricultural Sciences (BHU); 2021, 2021

[18] Hayes M, Svoboda M, Wall N, Widhalm M. The Lincoln Declaration on Drought Indices: Universal Meteorological Drought Index Recommended. Bulletin of the American Meteorological Society. 2011;**92**(4): 485-488. DOI: 10.1175/2010BAMS3103.1

[19] McKee TB, Doesken NJ, Kleist J. The Relationship of Drought Frequency and Duration to Time Scales. Boston, MA: Proceedings of the 8th Conference on Applied Climatology, 17–22 January 1993, Anaheim, CA; American Meteorological Society; 1993

[20] Available from: https://www.droughtmanagement.info/

[21] Hunt E, Hubbard K, Wilhite D, Arkebauer T, Dutcher A. The development and evaluation of a soil moisture index. International Journal Climatology. 2008;**29**:747-759. DOI: 10.1002/joc.1749

[22] Christov ID. Comparison of new computerized decision support eco-technology for precise irrigation scheduling with available products on market with similar function. Journal of Balkan Ecology. 2014;**17**(2):17-27

[23] Leonov EA, Leonova NE. Results of studying the water regime of soils and small catchments in the Don river basin. Vol. 194. Proceedings of the GGI; 1972. pp. 133-154 (in Russian)

[24] Terres J-M, Toth T, Wania A, Hagyo A, Koeble R, Nisini L. Technical report - Updated Guidelines for Applying Common Criteria to Identify Agricultural Areas with Natural Constraints. JRC; 2016

[25] Doorenbos, Pruitt. Crop water requirements. FAO Irrigation and Drainage Paper No. 24. Rome: Food and Agricultural Organization of the U.N. Rome; 1977

[26] Heim RR. A review of twentieth-century drought indices used in United States. BAMS. 2002;**83**:1149-1165

[27] Allen RG, Smith M, Perrier A, Pereira LS. An update for the definition of reference evapotranspiration. ICID Bulletin. 1994;**43**(2):1-34

[28] Allen GR et al. Crop Evapotranspiration-Guidelines for Computing Crop Water Requirements-FAO Irrigation and Drainage Paper 56. FAO; 1998. p. 300

Chapter 3

Modeling the Effect of Climate Change on Water Stored above a Micro-Dam in an Inland Valley Swamp in Sierra Leone, Using SWAT

Mohamed M. Blango, Richard A. Cooke, Juana P. Moiwo and Emmanuel Kangoma

Abstract

Many societies have experienced water scarcity resulting from population growth, increased urbanization and industrialization, increased irrigation associated with advances in agriculture productivity, desertification, global warming, or poor water quality. Climate change, and soil heterogeneity has a direct impact on the discharges of many rivers in and around the world. Various hydrological models have been used to characterize the impact of climate and soil properties on hydrology and water resources. The SWAT (Soil and Water Assessment Tool) water balance model, one such model, has been used at a variety of scales. In this instance it was used to model the impact of climate change on water storage in a reservoir at the downstream end of a small (75 ha) watershed. The watershed is the major component of an inland valley swamp, with a valley bottom that receives runoff from the watershed. The SWAT model was calibrated using storage data from 2014/15 and validated with data from 2015/16. Using future ensemble values derived from GCMs, the model predicted a reduction in the storage volume at the beginning of December of every dry season, with the 100-year storage volume down from 10,000 to 6900 cubic meters.

Keywords: climate change, micro-dam, inland valley swamp

1. Introduction

Advances in economies and standards of living have resulted in a growing dependency on water resources across the globe. Many societies have experienced water scarcity resulting from population growth, increased urbanization and industrialization, increased energy use, increased irrigation associated with advances in agriculture productivity, desertification, global warming, and poor water quality. Improved understanding of how each of these factors influences water supply, demand, and quality requires an understanding of the underlying processes, and their impact, on water availability and use. Such an understanding is best facilitated by a holistic approach that integrates hydrologic processes at the watershed scale, to determine an overall watershed response to user demands, changing climates, or both [1].

The circulation of water within the earth and atmosphere is a complex interaction of energy exchange and transportation pathways [2]. Hydrology addresses the waters of the earth, their occurrence, circulation and distribution, their chemical and physical properties, and their reaction with the environment, including their relation to living things. It also deals with the relationship of water with the environment within each phase of the hydrologic cycle. Factors such as rapid urbanization, industrialization and deforestation, land cover change, irrigation, have precipitated changes in hydrologic systems. Climate change, and soil heterogeneity also has a direct impact on the discharges of many rivers. Different hydrologic phenomena and hydrologic cycles need to be thoroughly studied to characterize these impacts.

Various hydrological models have been developed to characterize the impact of climate and soil properties on hydrology and water resources. Each model has its own unique characteristics, but common inputs used by deterministic models include rainfall, air temperature, soil characteristics, topography, vegetation, hydrogeology and other physical parameters. These models can be applied in very complex and large basins [3]. They have been used to simulate water movement under different conditions. They are used to study, *inter alia*, the impact of climate change on water availability, land use change on river discharges, and agricultural management strategies on water availability and sediment yield [2]. Central to this effort, watershed modeling is being utilized as a tool to better understand surface and subsurface water movement and the interactions between these water bodies. More importantly, they offer tools to guide decision making on water resources, water quality, and related hazard issues [1].

Models are a type of tool and are used in combination with many other assessment techniques. Models reflect how watershed systems are conceived. As with any tool, the answers they give are dependent on how they are applied, and the quality of these answers reflects how well the systems are understood. Models are also useful for extrapolating from current conditions to potential future conditions. Indeed, it is not possible to monitor the future, so modeling is the default choice. The ability of models to predict future conditions is very useful for projecting the outcomes of various possible management measures and strategies (http://www.epa.gov/watertrain).

The effect of climate change on water resources can be best handled through simulation of the projected hydrological conditions under such change. Such a treatment is essential as hydrological response is a highly complex process governed by a large number of variables such as terrain, landuse, soil characteristics and the state of the moisture in the soil. The SWAT (Soil and Water Assessment Tool) water balance model is one such model that has been used to carry out the hydrologic modeling of river basins [4].

SWAT is a computationally efficient simulator of hydrology and water quality at various scales. It is a mechanistic continuous model that can handle large watersheds in a computationally efficient manner. The "Hydrologic Unit Model for the United States" (HUMUS) has already been used to simulate river discharges at approximately 6000 gauging stations in the United States, with good results [5–7]. This study was extended within the national assessment of the USDA Conservation Effect Assessment Project (CEAP, (http://www.nrcs.usda.gov/Technical/nri/ceap/ceapgeneralfact.pdf.). SWAT is a comprehensive model that requires a diverse set of input parameters. However, many of the input parameters that are used to simulate special features are not available for all watersheds. SWAT was developed to quantify the impact of land management practices on water, sediment, and agricultural chemical yields in large complex watersheds with varying soils, land use, and management conditions, over long periods of time. The main components of SWAT are hydrology, climate, nutrient cycling, soil temperature, sediment movement, crop growth, agricultural management, and pesticide dynamics [8].

Terink et al. [2] described SWAT as a distributed rainfall-runoff model that divides a catchment into smaller discrete calculation units for which the spatial variation of the major physical properties are limited, and hydrological processes can be treated as being homogeneous. The total catchment behavior is a net result of many sub-catchments. Soil and land cover data are used to generate unique combinations within each sub-catchment, and each combination is considered as a homogeneous physical entity, called a Hydrological Response Unit (HRU). The water balance for HRUs is computed on a daily time basis. Hence, SWAT disaggregates the river basin into units that have similar characteristics in terms of soil and land cover, and that are located in the same sub-catchment. SWAT is capable of predicting the effects of changes in climate and management conditions on basin hydrology. The model includes many modules, including crop growth, groundwater, and river routing the facilitate the required simulations.

The model is semi-distributed, dividing each subbasin into HRUs based on soil type, crop patterns, and management practices [9]. SWAT has produced favorable model results when evaluated on watersheds with a range of conditions in the U.S. and other countries as diverse as Morocco [2], Finland [10], Pakistan [11] and India [4]. Across many of these watersheds, SWAT has shown flexibility in simulating surface runoff. Gassman et al. [12] found that, because of its core strengths, SWAT is primarily used for calibration and/or sensitivity analysis, climate change impacts, GIS interface descriptions, hydrologic assessments, variation in configuration or data input effects, comparisons with other models or techniques, interfaces with other models, and pollutant assessments. The foundation strength of SWAT in many of these applications is the combination of simplified upland and channel processes that are incorporated into the model [12]. The incorporation of the curve number (CN) method and non-spatial HRUs supports model adaptation to virtually any watershed with a wide variety of hydrologic conditions. With a flexible framework, SWAT facilitates the calculation of Total Maximum Daily Loads (TMDLs) and simulation of a wide variety of conservation practices and other BMPs (e.g., fertilizer and manure application rate and timing, irrigation management, and flood prevention structures). Several studies illustrate the success of SWAT in conducting these types of simulations as part of an overall BMP assessment [1].

The object of this study was to use SWAT to determine the effect of projected climate change on water storage and availability in a small reservoir developed by damming the upper reaches of an inland valley swamp in Sierra Leone. The reservoir is used to grow a second rice crop that extends into the dry season. Information on storage can be used to determine irrigable area under different climate change scenarios.

2. Methodology

2.1 Description of study area

The study area is a small 75 ha watershed, located at Njala University, in the Southern Province of Sierra Leone. The watershed is a major component of an inland valley swamp, with a valley bottom that receives runoff from the watershed. In a typical swamp, the stream formed in the valley bottom may be perennial, or ephemeral.

The watershed lies between latitudes 8.1140° and 8.1225° North and longitudes 12.0591° and 12.0668° west (**Figure 1**). The mean annual rainfall at Njala is 2200 mm. The annual Reference Evapotranspiration in 2015 and 2016 were 1322.3 and 1395.1 mm respectively.

There are two soils types in the watershed, Pelawahun loam and Njala sloping. Pelewahun soils occur in the drainageways of inland swamps that dissect the higher regions made up of Njala, Mokonde, and Bonjema soils. The terrain is nearly level to depressional, with concave slopes of 1 to 3 percent. Textures are loam to fine sandy loam in the topsoil, changing into clay loam in the upper subsoil, gravelly clay loam in the lower subsoil, and, as the bedrock is approached, clay [13].

2.2 SWAT model setup

SWAT requires data on terrain, land use, soil and weather for assessment of water yield at specified locations in a drainage basin. The selection of scale for spatial data is based on the tradeoff between the availability of required terrain data (in the form of contours data on the topographic maps) and the processing effort required for its

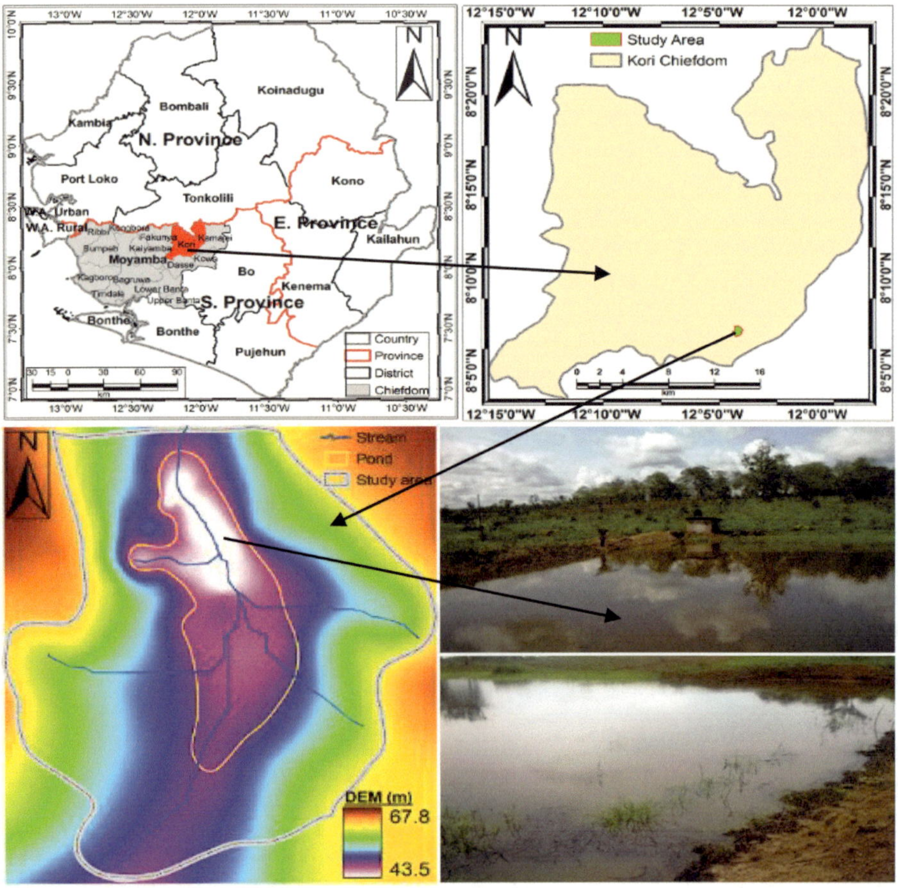

Figure 1.
Map of study area.

preprocessing using the GIS interface. The following sections provide brief description of data elements used and preprocessing performed on them.

2.2.1 Digital elevation model

A digital elevation model (DEM) is a topographic surface made up of elevation values laid out in a regular grid pattern. The DEM was generated using data from a topographic survey conducted over the watershed. The survey was carried out with the aid of Total station. The x-y-z data were then processed with the Quantum GIS (QGIS) into the DEM. QGIS is a public domain GIS that is free, easy to use, and available for Windows, Apple, and Android computers (http://qgis.org/en/site/). It also has the ability to convert maps between coordinate reference systems on the fly.

2.2.2 Delineation of the river basins

Automatic delineation of the river basins is done by specifying the location of the required final outflow point on the DEM as the final pour/drainage point. The watershed was divided into sub-basins using an arbitrarily selected threshold value. The number of sub basins is a function of this threshold, that controls the drainage density of the automatically constructed drainage system and thereby, the number of sub-basins. The DEM along with the main channels are shown in **Figure 2.**

2.2.3 Land cover/land use layer

Maps that outlined the landuses in the watershed were produced. These maps were stored as shapes files in QGIS, and converted to Comma Separated Variable file (.csv) files for input into SWAT. The entire watershed was split into four different landuses:

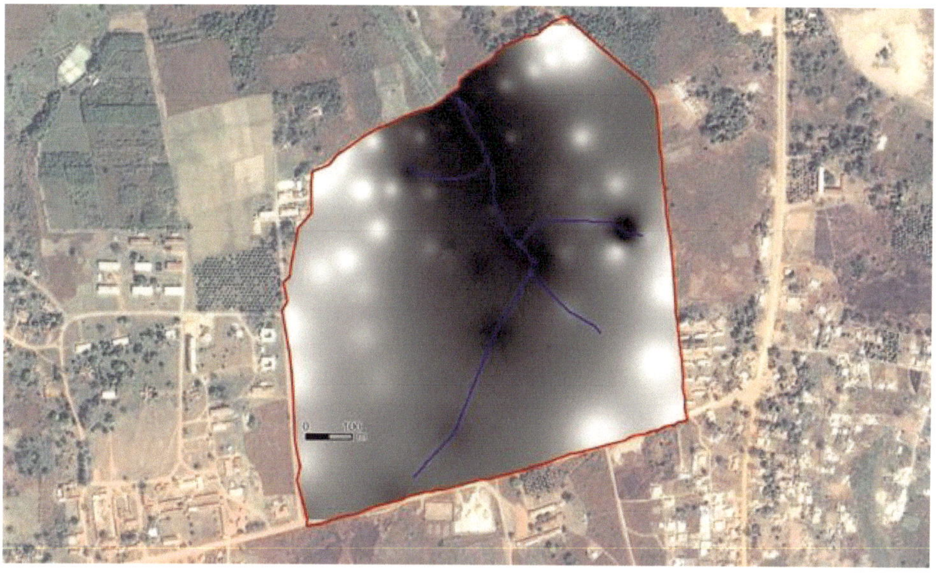

Figure 2.
Digital elevation model (DEM) with drainage network.

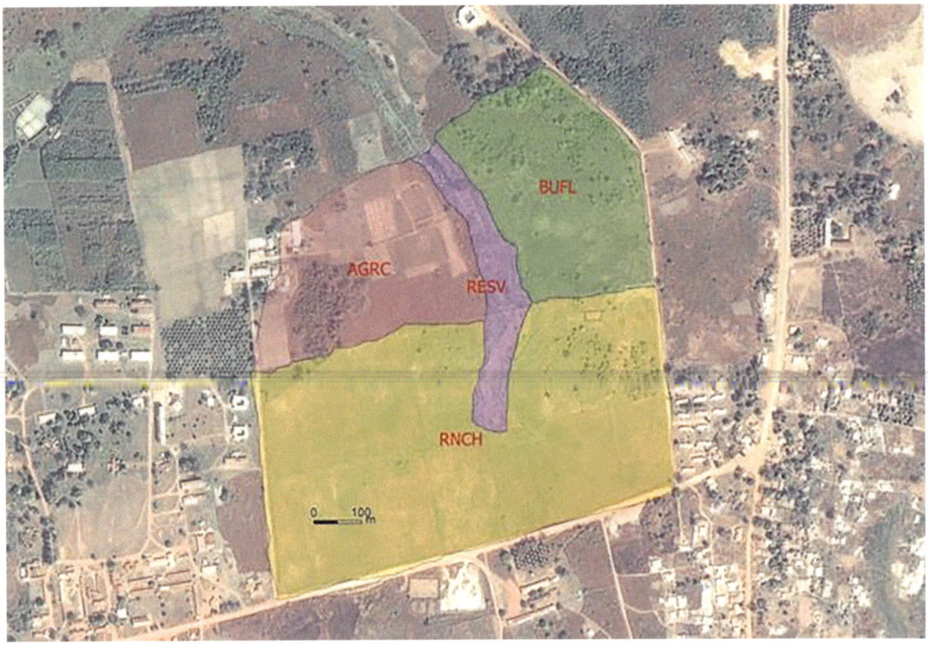

Figure 3.
Land use map for the watershed.

Grass range (RNCH), mixed forest (BUFL), Water (WATR) and agricultural row crops (AGRC). The land use map for the watershed is shown in **Figure 3.**

2.2.4 Soil layer

A shape file with the various soils in the watershed was also produced. Input tables were prepared, describing various parameters of the soil. These properties include bulk density, particle size distribution, water holding capacity, depth and number of layers of the soil. Two main types of soil were identified in the watershed; Njala sloping (Orthoxic Palehumult) and the Pelewahun loam (Typic Plinthaquult). The soil map is shown in **Figure 4**.

2.2.5 Weather data input preparation

Daily meteorological data were obtained from a WatchDog Weather Station (the 2009 ET Model, Spectrum Technologies Inc., USA) located next to the study area.

The weather input variables required included: solar radiation, wind speed, relative humidity, and precipitation, minimum and maximum temperatures. Weather data from November 2014 through January 2017 were used to calibrate and validate the model. Average data from nine climate change models were used to predict the change in stored water for mid-century (2046–2064) and the end of century (2081–2100).

2.2.6 Simulated volume of water in the pond

In SWAT, ponds and wetlands are impoundments that are located within the subbasin area. The .pnd file contains parameter information used to model the water,

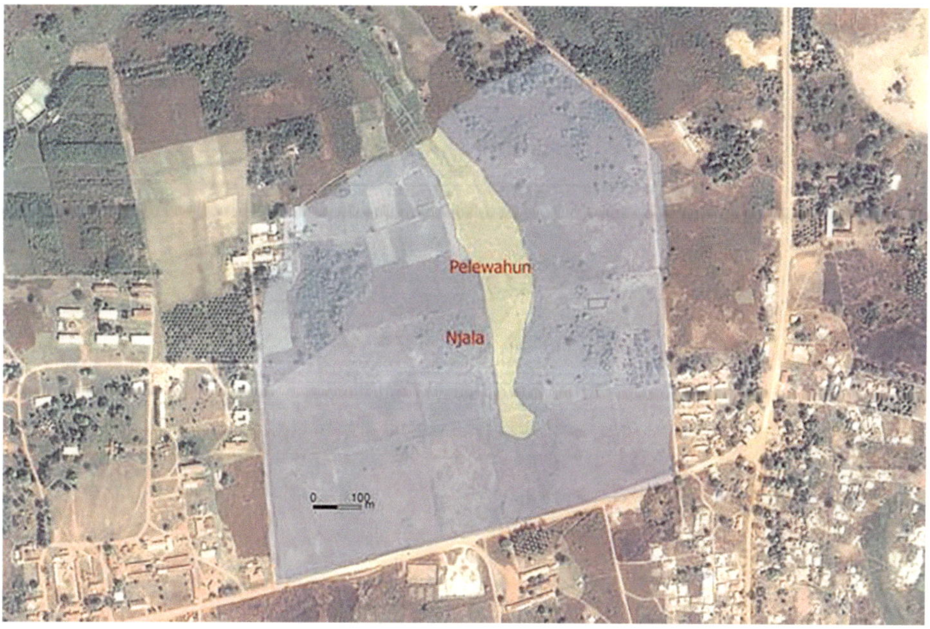

Figure 4.
Soil map for the watershed.

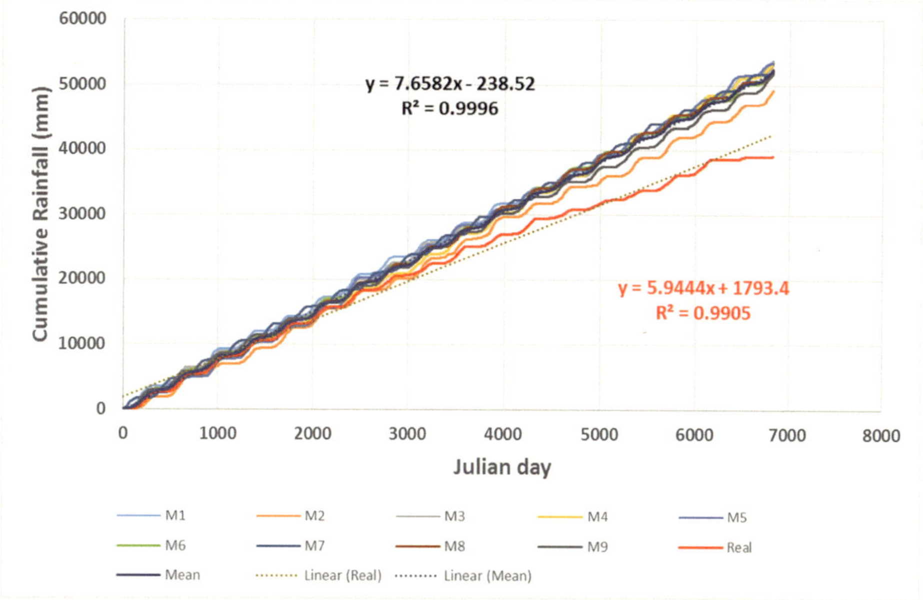

Figure 5.
Comparison of 20 years of simulated and historical rainfall for a 20-year period in Sierra Leone.

sediment and nutrient balance for ponds and wetlands. The key variables in the subbasin pond input file include: surface area (ha) and volume of water (m^3) in ponds when filled to principal spillway (ha), surface area and volume of water stored in ponds when filled to the principal spillway.

2.2.7 Future climatic data

For evaluation of the impact of climate change on water quantity, future climate (monthly precipitation and temperature) data were obtained from Bias Corrected and Downscaled WCRP CMIP3 Climate Projections archive for 2046–2064 and 2081–2100 (http://gdo-dcp.ucllnl.org/downscaled_cmip3_projections/). Weather and seasonal forecasts show improved reliability and consistency, when the outputs from multiple models are combined. Hence, the averages of nine GCM projections were used (ensemble) to drive the SWAT model in this study, rather than using the outputs from a single GCM.

The models were corrected for bias by comparing data predicted by the models for 1980 to 2000, with observed data from a station in Sierra Leone (**Figure 5**). Mid-century and end of century data produced by the models were then corrected using the ratio of the slopes.

3. Result and discussion

3.1 Pond output data

The HRU impoundment output file (.wtr) contains summary information for ponds, wetlands and depressional/impounded areas in the HRUs. The file is written in spreadsheet format. The output data include: precipitation falling directly on the pond (PNDPCP) during the time step (mm), the depth of surface runoff entering the pond (PND_IN) during the time step, evaporation (PNDEVP) from the pond surface during the time step (mm) and volume of water (m^3) in the pond (PNDVOL) at the end of time step.

3.2 Observed volume of water in the pond

The stage versus volume of water in the pond plot was used to determine the amount of water in pond at any point in time. A polynomial equation was used to model the relationship between stage (m) and volume of water. The equation was:

$$V = 9791\ h^2 + 25.818\ h - 119.05\ (R^2 = 0.997)$$

Where,
V = Volume of water in pond (m^3)
h = stage (m)

3.3 Coefficient of determination (R2) and Nash- Sutcliffe Efficiency (NSE)

The coefficient of determination (R^2) for the simulated and observed data of the volume of water in the pond was determined. The NSE was calculated using Eq. (1)

$$NSE = 1 - \left[\frac{\sum_{i=1}^{n} \left(Y_i^{obs} - Y_i^{sim} \right)^2}{\sum_{i=1}^{n} \left(Y_i^{obs} - Y^{mean} \right)^2} \right] \tag{1}$$

Where, Y_i^{obs} is the *i*th observation for the constituent being evaluated, Y_i^{sim} is the *i*th simulated value for the constituent being evaluated, Y^{mean} *mean* is the mean of observed data for the constituent being evaluated, and *n* is the total number of observations.

3.4 Average rainfall for different periods

The monthly average precipitation totals and average temperatures watershed for the baseline period (1961–2000), mid-century period (2046–2064) and late-century period (2081–2100). The baseline period values were computed using observed data, while the future ensemble values were derived from the GCMs, and corrected for bias, using the ratio of the slopes shown in **Figure 5**. Baseline and projected climate data are shown in **Figure 6**.

3.5 Model calibration

The model was calibrated by minimizing the root mean squared deviation (RMSD) between observed and simulated reservoir storage during the 2014/2015 dry cropping season. The two main factors affecting storage during this period was seepage and evaporation from the reservoir. The most sensitive parameter for evaporation in the model was the PNDEVCOEFF; the ratio of reservoir evaporation to potential evaporation, and PND_K, the hydraulic conductivity through the bottom of the pond. Both of these parameters were varied over the allowable ranges in the model, and the results were used to develop RSMD contours. These contours are shown in **Figure 7.**

Only certain combinations of the two parameters make physical sense. Under most conditions, PNDEVCOEFF varies between 0.6 and 0.8, with 0.7 being typically used as a default value. In this instance the minina band could not be extended down to these levels, as the model imposes a constraint on the PND_K, the hydraulic conductivity

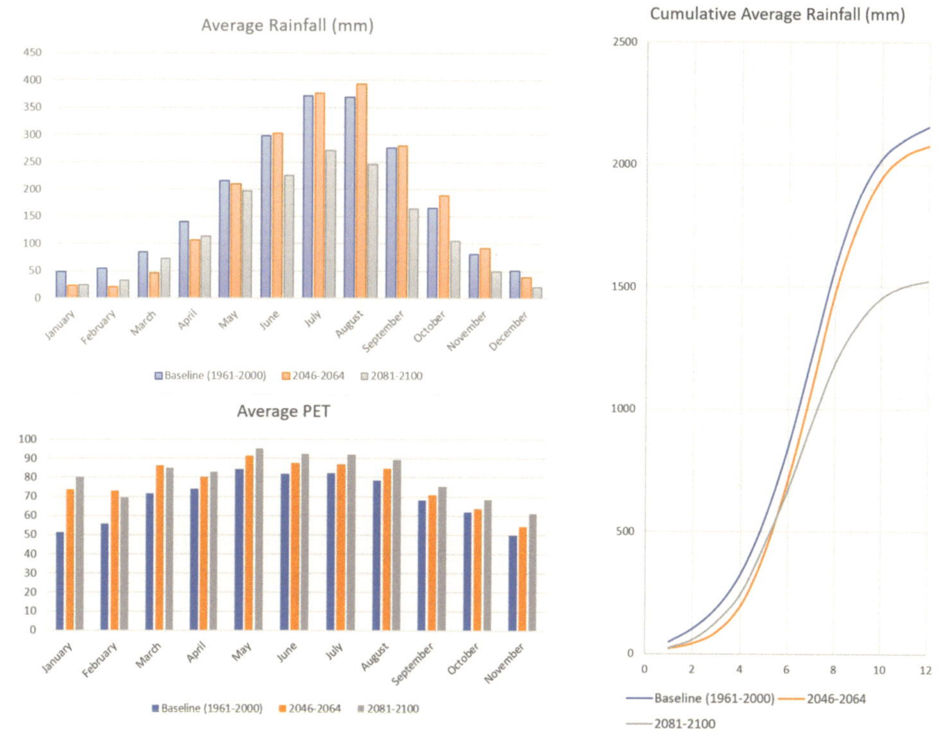

Figure 6.
Baseline and projected climate data at site.

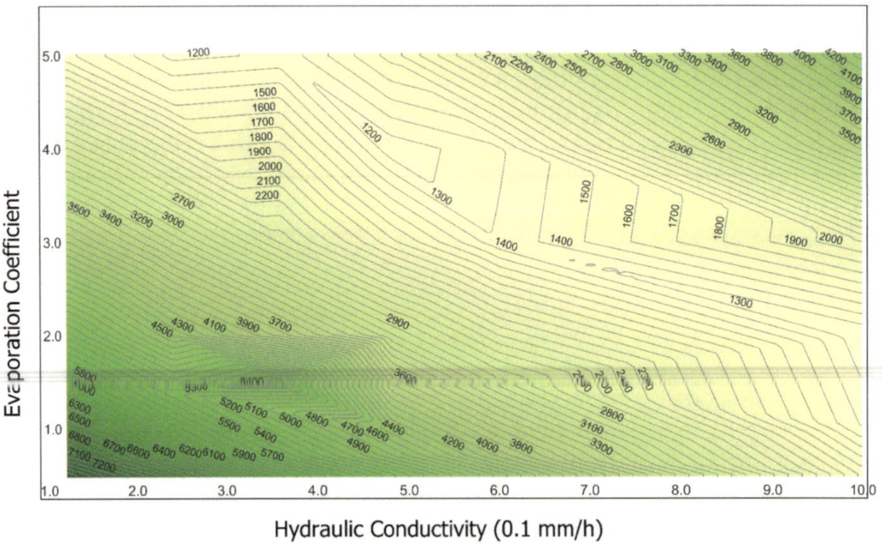

Figure 7.
RMSD contours of pond evaporation coefficient and hydraulic conductivity.

through the bottom of the reservoir. This maximum value is suitable for clay soils but is lower than values that are typical for loams, and sandy loams. SWAT was developed for US conditions where ponds are lined with a clay layer or some type of impermeable geotextile material. In this case, however, the bottom of the reservoir was unconsolidated Pelewahun loam. Given the constraints of the model, the calibrated value for PNDEVCOEFF was the value that gave the minimum RMSD value at the maximum allowable value of PND_K. The relationship between RMSD and PNDEVCOEFF when

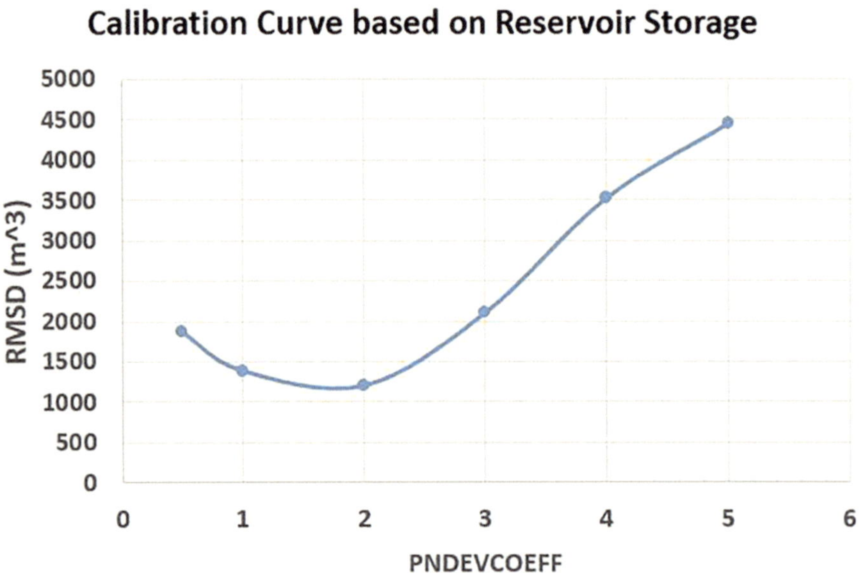

Figure 8.
RMSD versus pond evaporation coefficient.

PND_K is fixed at 1 mm/hr. is given in **Figure 8**. Based on this figure, 1 mm/hr. and 1.75 were used for PND_K and PNDEVCOEFF, respectively, in future simulations.

3.6 Model validation

Using the calibrated values of the coefficient of pond evaporation (PNDEVCOEFF) and hydraulic conductivity of pond (PND_K), the model was validated against the observed storage volume of 2015/16. The NSE and RMSE values were 0.82 and 1578.2 respectively.

3.7 Storage volume at the start of the dry season

The storage volumes at the start of December which typically marks the beginning of the dry season were fitted to cumulative distribution functions. The normal distribution was a better fit for 1981/2000, but the log normal better fitted the 2080/2099 period.

The baseline (1980–1999) and end of century periods (2081–2099) were utilized. The baseline data (1980–1999) had negative skewness (−0.09) and kurtosis of 1.22 and the end of century data had a skewness of 2.23 and a kurtosis of 5.89.

4. Discussion

The SWAT model has been found to be computationally efficient in its prediction and very useful because of its original design to assess the role of climate, topography,

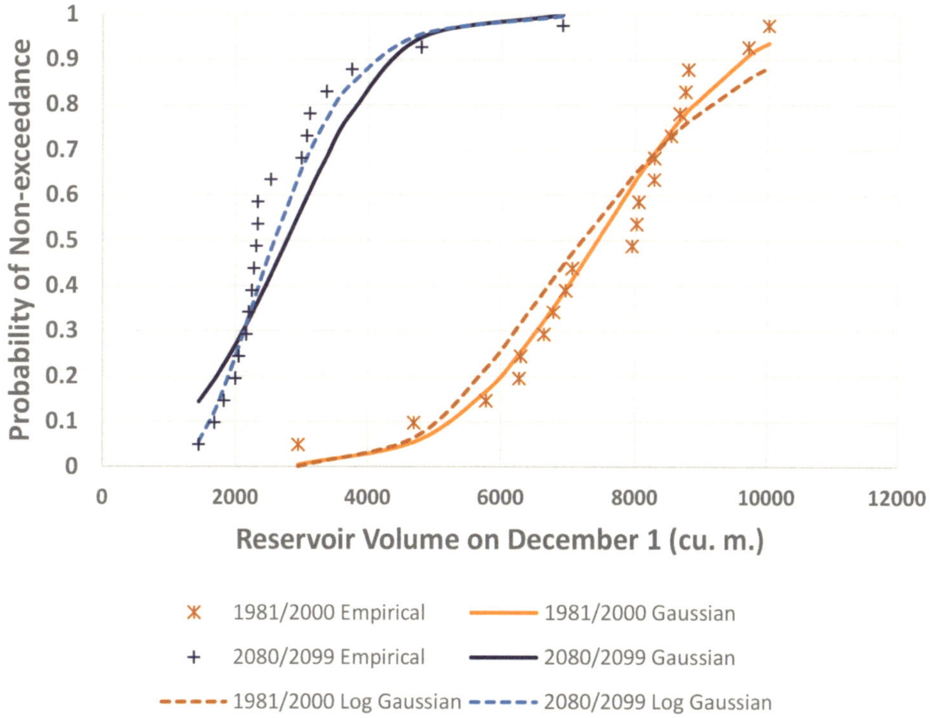

Figure 9.
Cumulative distribution functions for storage volume on December 1.

soil, and land use in the hydrologic response of a catchment [9, 14]. Several researches have shown satisfactory results in predicting impact of climate change on flows of streams, rivers and reservoirs [11, 15, 16].

In the calibration of the model, two parameters were considered: the coefficient of pond evaporation and the hydraulic conductivity (K). Although, in most cases K values range from 0.6 to 0.8, this parameter was calibrated at the maximum value of 1 for the model, as the model imposes a constraint on the hydraulic conductivity through the bottom of the reservoir. This could be attributed to the difference in soil texture in the bottom of ponds in the US where the model was developed as against the soil type in Sierra Leone for which the model is calibrated. Similarly, Beharry et al. [17] had to modify hydraulic conductivity as a key parameter in the calibration of the SWAT model.

Despite a high RMSE value for validation, the model also produced a good NSE value of 0.82. This indicates the model has a very high power of predicting the stored volume of water in the reservoir at any point in time. **Figure 9** in indicative of a reduction in storage volume of water in the reservoir at any probability level, in the future (2081–2100) as compared to the baseline period (1981–2000). This reduction could be attributed to an increase in evaporation due to high temperatures. This would have a negative impact- reducing the acres of land that can be irrigated. Subsequently, **Figure 9** also indicate that the 100-year storage volume will be reduced from a little over 10,000 m^3 during the baseline period to less than 8000 m^3 by the end of century (2081–2100). This would imply that designing storage for the dry season could involve less cost as storage volume is predicted to be reduced.

5. Conclusion

Using SWAT, the volume of water stored in a small reservoir was simulated. This was done considering the two main factors affecting storage during this period - seepage and evaporation from the reservoir. The most sensitive parameter for evaporation in the model was the ratio of reservoir evaporation to potential evaporation, and the hydraulic conductivity through the bottom of the pond. The model can be used to predict inflow into dams, water storage and release water management. For evaluation of the impact of climate change on water quantity, future climate (monthly precipitation and temperature) data were obtained from Bias Corrected and Downscaled WCRP CMIP3 Climate Projections archive for 2046–2064 and 2081–2100.

A knowledge of the expected storage volume in the reservoir at the start of the peak dry season is vital for proper planning of supplementary irrigation. Hydrologic modeling tools such as SWAT can help in predicting the hydrologic dynamics of watersheds. However, a challenge to utilizing such a tool will require a calibration process to determine the extent to which parameters can influence the predictive power of the model. The SWAT model was calibrated using the storage volumes of 2014/15 dry season and validated with the 2015/16 dry season data. With future ensemble values derived from the GCMs, the model predicted a reduction in the storage volume at the beginning of December of every dry season, with the 100-year storage volume down from 10,000 m^3 to 6900 m^3 (**Figure 9**).

The study highlighted the important role that future climate projections would play in understanding the effect of climate change on hydrologic processes. The use of models such as SWAT would be helpful in investigating land use change on nutrient loading and water quality in streams and rivers especially in sub-Saharan countries.

Author details

Mohamed M. Blango[1*], Richard A. Cooke[2], Juana P. Moiwo[1] and Emmanuel Kangoma[1]

1 Department of Agricultural Engineering, Njala University, Sierra Leone

2 Department of Agricultural and Biological Engineering, University of Illinois, Urbana-Champaign, USA

*Address all correspondence to: mmblango@njala.edu.sl

References

[1] Daniel EB, Camp JV, LeBoeuf EJ, Penrod JR, Dobbins JP, Abkowitz MD. Watershed modeling and its applications: A state-of-the-art review. The Open Hydrology Journal. 2011;**5**:26-50 1874-3781/Vanderbilt University, VU Station B 351831, Nashville, Tennessee 37235-1831, USA

[2] Terink W, Hunink J, Droogers P, Reuter H, van Lynden G, Kauffman S. Soil and Water Management Scenarios Using the Soil and Water Assessment Tool (SWAT). Morocco: Green Water Credits Morocco: Inception Phase. Green and Blue Water Resources for the Sebou Basin; 2011

[3] Devi GK, Ganasri BP, Dwarakish GS. A review on hydrological models. International conference on water resources, coastal and ocean engineering (ICWRCOE 2015). Aquatic Procedia. 2015;**4**(2015):1001-1007

[4] Gosain AK, Rao S, Basuray D. Climate change impact assessment on hydrology of Indian river basins. Special section: Climate change and India 346-53. Current Science. 2006;**90**(3):10

[5] Arnold JG, Srinivasan R, Muttiah RS, Allen PM. Continental scale simulation of the hydrologic balance. Journal of American Water Resources. Association. 1999;**35**(5):1037-1051

[6] Srinivasan RS, Arnold JG, Jones CA. Hydrologic modeling of the United States with the soil and water assessment tool. International Journal of Water Resources Development. 2010a;**14**(3):315-325

[7] Srinivasan R, Zhang X, Arnold J. SWAT ungauged: Hydrological budget and crop yield predictions in the upper Mississippi River basin. Transactions of the ASABE. 2010b;**53**(5):1533-1546 American Society of Agricultural and Biological Engineers ISSN 2151-0032

[8] Abbaspour KC, Faramarzi M, Rouholahnejad E. Hydrological Modeling of Alberta Using Swat Model. A Preliminary report; 2010. Available from: https://albertawater.com/dmdocuments/05_2010_12_Hydrological_Modeling_of_Alberta_Using_SWAT_Model%20(2)%20(7).pdf

[9] Neitsch SL, Arnold JG, Kiniry JR, Williams JR. Soil and water assessment tool theoretical documentation version. Grass land. Soil and Water Research Laboratory. 2005 Agricultural Research Service 808 East Blackland Road, Temple, Texas 76502; Blackland Research Centre, Texas Agricultural Experiment Station 720, East Blackland, Texas USA. Available from: https://swat.tamu.edu/media/1292/swat2005theory.pdf

[10] Bärlund I, Kirkkala T. Examining a model and assessing its performance in describing nutrient and sediment transport dynamics in a catchment in southwestern Finland. Boreal Environment Research. 2008;**13**:195-207

[11] Saifullah M, Adnan M, Zaman M, Walega A, Liu S, Khan MI, et al. Hydrological response of the Kunhar River basin, Pakistan to climate change and anthropogenic impacts on runoff characteristics. Water. 2021;**2021**(13):3163. DOI: 10.3390/w13223163

[12] Gassman PW, Reyes MR, Green CH, Arnold JG. The soil water assessment tool: Historical development, applications and future research directions. Transactions of the ASABE. 2007;**5**(4):1211-1250

[13] Odell RT, Dijkerman JC, van Vuure W, Melsted SW, Beavers AH, Sutton PM, et al. Characteristics, classification, & adaptation of soil in selected areas in Sierra Leone. In: West Africa Bulletin 748 Agricultural Experiment Station College of Agriculture. University of Illinois at Urbana-Champaign Bulletin 4 Njala University College. University of Sierra Leone. Agricultural Experiment Station, University of Illinois; 1974. Available from: https://agris.fao.org/agris-search/search.do?recordID=US201300528573

[14] Kibii JK, Kipkorir EC, Kosgei JR. Application of soil and water assessment tool (SWAT) to evaluate the impact of land use and climate variability on the Kaptagat Catchment River discharge. Sustainability. 2021;**13**:1802. DOI: 10.3390/su13041802

[15] Nazari-Sharabian M, Taheriyoun M, Ahmad S, Karakouzian M, Ahmadi A. Water quality modeling of Mahabad dam watershed–reservoir system under climate change conditions, using SWAT and system dynamics. Water. 2019;**11**(2):394. DOI: 10.3390/w11020394

[16] Saade J, Atieh M, Ghanimeh S, Golmohammadi G. Modeling impact of climate change on SurfaceWater availability using SWAT model in a semi-Arid Basin: Case of El Kalb River, Lebanon. Hydrology. 2021;**8**:134. DOI: 10.3390/hydrology8030134

[17] Beharry SL, Gabriels D, Lobo D. Use of the SWAT model for estimating reservoir volume in the upper Navet watershed in Trinidad. SN Applied Science. 2021;**3**:163. DOI: 10.1007/s42452-021-04201-7

Chapter 4

Exotic Vs. Autochthonous Grapevine Varieties – A Case Study on Global Warming in Northeastern Portugal

Manuel T. Oliveira and Ana A. Oliveira

Abstract

Grapevines, an economic mainstay of the Douro Demarcated Region, are under increasing stressful conditions and they can suffer further losses due to climate change. Observations on weather patterns and behavior of two autochthonous grapevines and two exotic ones were made over several years. There are indications of an increase of 2°C from 2003 to 2019 responsible for the advancement of 10 to 15 days of phenological events on all grape varieties, a clear biological sign of climate change. Against the forecasted trends, rainfall showed a trend for increasing total amount but a lower proportion during the growing season that resulted in stronger seasonality. The yields of native varieties were about 2600 kg ha^{-1} higher than exotic varieties, a difference supported by a larger leaf area, on average 1.7 m^2 higher, and better stomatal conductance in average 2.6 mm s^{-1} and 2.1 mm s^{-1} for native and exotic varieties, respectively. These differences suggest that natives are better suited to withstand aggravated environmental conditions than the exotic. The composition of the must show significantly higher total soluble content in autochthonous grapevines but they have a lower concentration of organic acids, tannins, and polyphenols, meaning poorer organoleptic profiles.

Keywords: Vitis vinifera L., phenology, yield components, warming temperature, climate change, stressful environment

1. Introduction

Viticulture is an economically important cropping system in many parts of the world and grapevines, a long-lived perennial is useful in bioclimatic studies as an indicator of both historical and contemporary climatic changes. The effect of temperature on grapevines and berry composition is well established and varieties are classified by their thermal requirements [1, 2].

The scientific community has largely accepted that the world climate is changing and the Mediterranean basin is experiencing a climate shift to become warmer and drier [3]. The Joint Research Center, the European Commission's science and knowledge service, estimates that the average temperatures in the Mediterranean region have risen by 1.4°C since the pre-industrial era, 0.4°C more than the global average and summer rainfall is at risk of being reduced by 10 to 30% in some regions [4]. The Portuguese territory has suffered a decreased precipitation since the 1950s and a significant upward shift in temperature [5, 6]. The Douro Demarcated Region (DDR), located in Northeastern Portugal is classified as a Denomination of Controlled Origin (DOC), the highest Portuguese wine classification, and it has an economic preeminence in the Portuguese wine industry. Few crops are as susceptible to minor changes in climate as vineyards, especially those grown for premium wine quality grapes [7], and any significant change in the Douro environment is likely to affect its economy and social conditions [8].

These trends in weather will affect the grapevine growth and the berry composition:

1. Yield losses due to berry shriveling and sunburn [9]

2. Earlier onset of phenological stages [10, 11]

3. Increment of sugar accumulation and consequent higher alcohol content in wine [9, 12]

4. Organic acids are metabolized faster and their final content in the must will be lower that in turn make the must microbiologically unstable [12]

5. The wine aroma profile is poorer towards overripe [13]

The major abiotic stresses that affect grapevine production in the Mediterranean region are drought, excessive light intensity, and heat [14] all of them present in DDR. Yield losses due to high solar radiation coupled with elevated temperatures are common in DDR and a shift in phenological dates have been already observed [15, 16]. Higher sugar content and low acidity are common characteristics of musts of the region where the addition of tartaric acid during winemaking is necessary to correct their low acidity [17–19].

Effects of climate on grape yield and quality are cultivar-dependent as different grape cultivars grown under common climate conditions still show large variations [20–22]. This diversity helps growers adapt wine grapes to shifting conditions, planting varieties better adapted to current and future climate regimes [23]. However, the introduction and spread of world-renowned varieties in many wine growing regions all over the world has caused a loss of indigenous grapevine varieties traditionally grown and left the farmers with a shrinking pool of genetic variability. In Portugal, there are identified 236 grape genotypes with origin in the country [24]. In 1920, 120 varieties were cultivated in DDR but actually, only 20 varieties dominate the vineyards of the region [25] and a few of them have an exotic origin. The widespread cultivation of world-renowned varieties has caused great losses of indigenous grapevine varieties traditionally grown [26]. Viticulture faces new challenges to respond to consumers' demands and particularly to climate change. Thus, characterization and preservation of this grapevine genetic background prospected, mainly of late-ripening cultivars, is of crucial importance to face alterations in temperature, precipitation, frequency and duration of extreme weather events, and also their resulting abiotic consequences [27].

Touriga Franca and Touriga Nacional are two Portuguese red cultivars very common in DDR were Cabernet Sauvignon and Syrah, two red cultivars of exotic origin, are also planted. Touriga Franca covers 22% of the planted area (IVDP, 2017), it is a hardy variety with good tolerance to heat, high yield and it can produce wines with complex, intense aromas [28]. Touriga Nacional covers about 8% of the planted area, it is also heat tolerant but has a lower yield. Both varieties are grown over a large range of thermal conditions [29] and are two of the most valuable premium varieties to produce quality wines with a unique aroma profile that can fetch high market prices [30, 31].

Cabernet-Sauvignon is now one of the most cultivated wine grapes in the world, about 5% of world vine area, and Syrah occupies about 2.5% [32]. In DDR they are dispersed and represent a small planted area but there is a tendency to increase it. In Australia, Cabernet Sauvignon and Shiraz are reported to be better suited to warmer climates as compared to other exotic varieties [33]. Cabernet Sauvignon tends to produce full-bodied wines with high tannins and noticeable acidity that contributes to the wine's aging potential. Syrah is consistently full-bodied with softer tannins.

During 4 years we followed the development, yield, and must characteristics of Touriga, Nacional, Touriga Franca, Cabernet Sauvignon, and Syrah that were planted just a few meters apart on rainfed plots. The field observations took place at the Eastern end of DDR where the climate conditions during the growing season are semi-arid, with high temperatures and intense solar radiation. In near future, the climate is forecasted to become drier and hotter, and the growers must have clear information on how to adjust their vineyards to face such projected scenarios [34]. The objective was to compare the agronomic performance of exotic to autochthonous grapevine varieties subjected to a stressful environment and predict which ones are better suited to withstand likely aggravated climate conditions.

2. Materials and methods

The vineyard is located at 41.148654 N, 7.127574 W (**Figure 1**), on a rainfed flat land with an Inceptisol Durixerept 70 to 90 cm deep [35]. The vines (V. vinifera L.) are grown in vertical shoot positioning and trellised as simple guyot. Each vine row is about 60 m long, 2 meters apart from each other, and 1 meter between plants of the same row. At full development, the rows formed a continuous hedge kept at a maximum of 170 cm high and 70 cm thick by mechanical trimming. The weeds were mowed between the rows and controlled by very shallow tillage within the row. Four grapevine varieties - Touriga Nacional (TN), Touriga Franca (TF), Cabernet Sauvignon (CS), Syrah (S) – occupy two selected rows on each plot.

An on-site meteorological station provided data from 2003 to the present at 5 minutes intervals averaged over 1 hour. During the same time period, the average dates for each phenological stage [36], computed over all varieties grown in that particular farmstead, were also recorded.

The experiment with the four mentioned varieties took place between 2016 and 2019 and from the start 40 plants of each variety were randomly chosen. Phenological stages [36] were recorded individually on each plant and it was calculated the average date each stage occurred for a given variety. Harvest date was determined by the usual commercial approach when the Brix degree of the must reached the maximum value. As soon as the clusters were visible, one cluster per plant was clearly marked to follow its development from flowering to harvest. At flowering, the number of flowers per cluster was counted using an artificial vision technique [37] and the number of

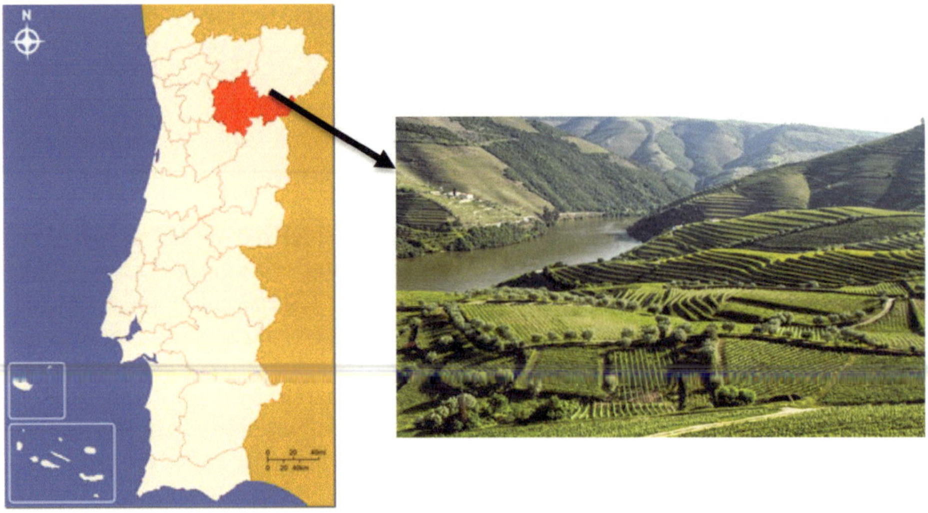

Figure 1.
DDR location in Northeast Portugal and the experimental site.

berries at harvest was counted manually excluding the shriveled and the sunburnt berries. Every 15 days from flowering to harvest, the total leaf area (tla) per plant was estimated using the reduction of solar radiation crossing the canopy ([38] and the stomatal conductance (g_s - mm s^{-1}) was measured (AP4 porometer, Delta-T Devices, www.delta-t.co.uk) at solar noon on clear sky days on one well-exposed adult leaf out of 10 of the 40 choose plants of each variety. From the clusters harvested, three samples of about 400 g each were set up for laboratory analysis of the must [39]: total soluble solids, titratable acidity, pH, glucose and fructose sugars, malic, and tartaric acids, tannins, polyphenols, and anthocyanins.

The thermal requirements to reach the phenological stages were expressed in Growing Degree Days (GDD), and agroclimatic indicator related to the growth cycle of plants, calculated as [11]:

$$GDD = \sum_{ti}^{tn} (T - Tb)$$

where ti is the starting day, tn the final day, T is the average daily temperature, and Tb is the base temperature. The commonly accepted standard Tb in viticulture literature is 10°C [40] a value that was considered adequate for calculating GDD in DDR [15].

The statistical layout was a completely randomized design with two main factors: variety (4) and year (4). Tukey HSD was used for mean separation. The analyses were performed with SPSS statistical package (SPSS for Windows release 20, SPSS Inc. 2011, Chicago).

3. Results

The records of the meteorological station show that air temperature is significantly different ($P \leq 0.05$) among the years (2003 to 2019) and a tendency for a steady rise. Average temperature increased about 2°C in 16 years, a mark that should be reached only sometime in 2050 according to [41]. The average temperature for the grapevine growing season, Abril to September, did not show a significant difference among years, suggesting that the annual increase is mostly dependent on higher winter temperatures. Nevertheless, the average temperatures of every growing season have been above 21°C, considered the upper limit for producing high-quality wines [42].

Annual precipitation shows a significant augmenting trend that is related to winter precipitation as rainfall during the growing season has shown no significant differences among seasons. However, the number of rainy days (over 1 mm day^{-1}) per year has decreased year after year, suggesting higher rainfall intensities per rain event, a phenomenon that favors runoff at expense of water infiltration and soil water storage that might aggravate water shortages during the drier periods and soil erosion. The rainfall did not follow the prediction for the Mediterranean areas [41] but the lower proportion of precipitation falling during the growing season translates into stronger seasonality [43].

The budbreak (b) took place 70 to 74 days after 1 January (average 71 days) with no significant (P > 0.05) differences among varieties and years (**Tables 1** and **2**). Flowering (f) occurred about 10 days earlier for CS and S than for TN and TF. After

Variety	Year	1 Jan to budbreak		Budbreak to flowering		Flowering to veraison	
		d	GDD	D	GDD	d	GDD
CS	2016	69^{a}	67.8^{a}	77^{a}	300.8^{a}	43^{a}	547.5^{a}
S		70^{a}	67.8^{a}	79^{a}	318.9^{a}	42^{a}	567.0^{a}
TN		69^{a}	67.8^{a}	84^{b}	350.4^{b}	44^{a}	625.9^{b}
TF		69^{a}	67.8^{a}	86^{b}	361.9^{b}	41^{a}	586.8^{b}
CS	2017	71^{a}	68.2^{a}	65^{a}	398.5^{a}	51^{a}	911.5^{a}
S		71^{a}	68.2^{a}	67^{a}	415.9^{a}	51^{a}	935.1^{a}
TN		69^{a}	68.1^{a}	76^{b}	484.0^{b}	53^{a}	1005.9^{b}
TF		69^{a}	68.1^{a}	79^{b}	560.5c	51^{a}	976.4^{b}
CS	2018	72^{a}	66.3^{a}	82^{a}	412.5^{a}	43^{a}	551.4^{a}
S		73^{a}	68.9^{a}	82^{a}	416.8^{a}	40^{a}	516.7^{a}
TN		69^{a}	64.1^{a}	94^{b}	481.4^{b}	42^{a}	609.0^{b}
TF		70^{a}	65.9^{a}	91^{b}	459.6^{b}	45^{a}	615.1^{b}
CS	2019	70^{a}	67.4^{a}	84^{a}	417.9^{a}	45^{a}	561.3^{a}
S		73^{a}	67.1^{a}	81^{a}	420.1^{a}	41^{a}	527.6^{a}
TN		70^{a}	65.5^{a}	95^{b}	472.3^{b}	41^{a}	600.1^{b}
TF		70^{a}	64.8^{a}	94^{b}	465.1^{b}	44^{a}	620.2^{b}
	2016	69^{a}	67.8^{a}	75^{a}	333.0a	43^{a}	581.8^{a}
	2017	70^{a}	68.2^{a}	72^{a}	464.7^{b}	42^{a}	557.2^{a}
	2018	71^{a}	66.3^{a}	87^{b}	442.6^{b}	43^{a}	773.1^{b}
	2019	71^{a}	66.9^{a}	88^{b}	470.3^{b}	44^{a}	748.1^{b}

Different superscript letters on the same column and pertaining to varieties grouped by years mean significant difference (Tukey's $HSD_{0.05}$).

Table 1.
Number of days (d) and growing degree days (GDD) to reach a phenological stage until veraison across cabernet sauvignon, Syrah, Touriga Nacional, and Touriga Franca from 2016 to 2019.

Variety	Year	Veraison to harvest		1 Jan to harvest	
		D	GDD	d	GDD
CS	2016	52^{a}	924.3^{a}	241^{a}	1840.5^{a}
S		49^{a}	871.2^{b}	240^{a}	1824.9^{a}
TN		53^{a}	956.0c	250^{b}	2000.1^{b}
TF		51^{a}	921.5^{a}	247^{b}	1938.0^{b}
CS	2017	51^{a}	804.1^{a}	238^{a}	2182.2^{a}
S		52^{a}	818.0^{a}	241^{a}	2237.2^{b}
TN		47^{a}	721.9^{b}	245^{b}	2279.9^{b}
TF		49^{a}	747.0^{b}	248^{b}	2351.9^{c}
CS	2018	49^{a}	805.9^{a}	246^{a}	1836.1^{a}
S		51^{a}	834.1^{a}	246^{a}	1836.5^{a}
TN		48^{a}	779.2^{b}	253^{b}	1933.8^{b}
TF		48^{a}	777.7^{b}	254^{b}	1918.3^{b}
CS	2019	50^{a}	825.3^{a}	249^{a}	1871.9^{a}
S		50^{a}	825.2^{a}	245^{a}	1840.0^{a}
TN		51^{a}	852.1b	257^{b}	1990.0^{b}
TF		49^{a}	849.7b	257^{b}	1999.8^{b}
	2016	51^{a}	718.3^{a}	238^{a}	1700.9^{a}
	2017	50^{a}	772.7^{b}	234^{a}	1862.8^{a}
	2018	49^{a}	799.2^{b}	250^{b}	2081.2^{b}
	2019	52^{a}	854.3^{b}	255^{b}	2139.6^{b}

Different superscript letters on the same column and pertaining to varieties grouped by years mean significant difference (Tukey's $HSD_{0.05}$).

Table 2.
Number of days (d) and growing degree days (GDD) to reach the harvest across cabernet sauvignon, Syrah, Touriga Nacional, and Touriga Franca from 2016 to 2019.

flowering, the length of time to reach any of the next stages was about the same for all varieties. After budbreak, all phenological stages occurred later in 2018 and earlier in 2017 with no significant interaction between year and variety. The longer period of time necessary for TN and TF to reach f and the coincidence of the later stages with periods of warmer weather was conducing to about 100 GDD in excess to S and CS at harvest.

The number of clusters per vine varied between 25 and 28 (average 26) independently of the variety or production year. The vine load is determined by the winter pruning and it was kept at the same level every year as usual commercial practice. The number of flowers per cluster is a variety characteristic (**Table 3**) and the native varieties TN and TF usually had significantly more flowers per bunch than the exotic ones. A larger proportion of the flowers present at f resulted in a larger number of mature berries for TN and TF than for CS and S. One consequence of a larger number of berries per bunch at harvest is a higher average weight per bunch and increased productivity.

Variety	Year	Num. flowers per cluster	Num. berries per cluster	Avg. weight per cluster (g)	Yield (kg ha^{-1})
CS	2016	204.4^a	115.5^a	140.3^a	3416.8
S		182.0^b	92.6^a	110.5^b	2742.1
TN		259.4^c	186.9^b	176.4^c	5049.7
TF		254.1^c	174.1^b	171.7^c	4560.6
CS	2017	123.0^a	41.8^a	51.2^a	1338.5
S		72.5^b	55.8^a	117.6^b	2976.1
TN		156.3^c	91.8^b	139.2^c	4227.6
TF		115.7^c	75.2^c	133.8^c	3846.0
CS	2018	116.9^a	86.5^a	104.6^a	2558.9
S		72.9^b	33.7^b	41.3^b	970.9
TN		162.1^c	146.9^c	222.5^c	6410.2
TF		202.3^d	160.4^d	241.5^c	5982.5
CS	2019	130.2	57.1	88.7	2335.0
S		85.4	48.3	52.6	2178.1
TN		170.8	151.8	227.8	5762.5
TF		198.2	181.6	190.3	4098.4
	2016	250.0^a	109.8^a	149.3^a	3942.3
	2017	119.0^b	81.2^b	110.5^b	2847.1
	2018	129.4^b	106.9^a	152.5^a	3980.6
	2019	131.8^b	115.7^a	161.8^a	4021.9

Different superscript letters on the same column and pertaining to varieties grouped by years mean significant difference (Tukey's $HSD_{0.05}$).

Table 3.
Mean separation of a number of flowers at flowering and berries at harvest per cluster, cluster weight, and yield per hectare (extrapolated average) across cabernet sauvignon, Syrah, Touriga Nacional, and Touriga Franca from 2016 to 2019.

The higher productivity of native varieties is supported by a larger canopy, more leaf area in total (**Tables 4** and **5**), and a higher stomatal conductance that might increase the photosynthetic rate. The canopies reached their largest development a few days after veraison, 4 to 5 m^2 for TN and TF and 2 to 3 m^2 for CS and S. Stomatal conductance decreased from flowering to harvest ranging from 3.47 to 1.47 mm s^{-1}.

The composition of the must show significant differences between the native and the exotic varieties (**Tables 6** and **7**). TF and TN had higher total soluble content (higher °Brix) that is related to higher sugar (glucose and fructose) concentration. On other hand, the concentration of organic acids (malic and tartaric) and of tannins and polyphenols are higher in must of S and CS. Titrable acidity and pH showed no significant differences among varieties or they were of no enological significance. It was found significant differences in total anthocyanin but no clear distribution among the varieties.

Variety	Year	Flowering		Midterm		Veraison	
		tla (m^2)	g_s (mm s^{-1})	tla (m^2)	g_s (mm s^{-1})	tla (m^2)	g_s (mm s^{-1})
CS	2016	2.48^{a}	3.01^{a}	2.61^{a}	2.59^{a}	2.74^{a}	2.11^{a}
S		2.24^{a}	3.04^{a}	2.35^{a}	2.68^{a}	2.47^{a}	2.08^{a}
TN		4.40^{b}	3.99^{b}	4.75^{b}	2.87^{b}	5.11^{b}	2.56^{b}
TF		3.40^{c}	3.87^{b}	3.92^{c}	2.98^{b}	4.06^{b}	2.68^{b}
CS	2017	2.21^{a}	3.11^{a}	2.42^{a}	2.87^{a}	2.62^{a}	2.54^{a}
S		2.10^{a}	3.13^{a}	2.38^{a}	2.97^{a}	2.45^{a}	2.50^{a}
TN		3.52^{b}	3.88^{b}	3.89^{b}	3.02^{a}	4.86^{b}	2.80^{b}
TF		2.65^{b}	3.85^{b}	3.09^{c}	3.24^{b}	3.71^{c}	2.91^{b}
CS	2018	2.59^{a}	3.20^{a}	3.03^{a}	2.79^{a}	3.38^{a}	1.82^{a}
S		2.75^{a}	3.35^{a}	2.74^{b}	2.87^{a}	2.93^{b}	1.72^{a}
TN		4.85^{b}	3.57^{b}	4.87^{c}	2.99^{a}	5.37^{c}	2.56^{b}
TF		3.96^{c}	3.63^{b}	4.05^{d}	2.86^{a}	4.11	2.56^{b}
CS	2019	2.40^{a}	3.21^{a}	2.54^{a}	2.64^{a}	2.78^{a}	2.06^{a}
S		2.63^{a}	3.15^{a}	2.61^{a}	2.51^{a}	2.80^{a}	2.15^{a}
TN		4.06^{b}	3.62^{b}	4.29^{b}	2.89^{b}	4.41^{b}	2.55^{b}
TF		4.14^{b}	3.58^{b}	3.87^{b}	3.10^{b}	3.87^{b}	2.81^{b}
	2016	3.13^{a}	3.48^{a}	3.34^{a}	2.78^{a}	3.56^{a}	2.36^{a}
	2017	2.62^{b}	3.49^{a}	2.94^{b}	3.03^{b}	3.41^{a}	2.69^{b}
	2018	3.54^{c}	3.44^{a}	3.67^{c}	2.88^{a}	3.95^{b}	2.14^{c}
	2019	3.45^{c}	3.41^{a}	3.14^{d}	2.99^{a}	3.74^{b}	2.29^{a}

Different superscript letters on the same column and pertaining to varieties grouped by years mean significant difference (Tukey's $HSD_{0.05}$).

Table 4.
Mean separation of total leaf area (tla) and stomatal conductance (g_s) from flowering to veraison across cabernet sauvignon, Syrah, Touriga Nacional, and Touriga Franca from 2016 to 2019.

Variety	Year	Midterm		Harvest	
		tla (m^2)	g_s (mm s^{-1})	tla (m^2)	g_s (mm s^{-1})
CS	2016	2.80^{a}	$1{,}56^{a}$	2.80^{a}	1.21^{a}
S		2.50^{a}	1.55^{a}	2.50^{a}	1.23^{a}
TN		5.31^{b}	1.86^{b}	5.36^{b}	1.59^{b}
TF		4.06^{b}	1.78^{b}	4.07^{c}	1.50^{b}
CS	2017	2.54^{a}	1.59^{a}	2.35^{a}	1.32^{a}
S		2.40^{a}	1.68^{a}	2.16^{a}	1.40^{a}
TN		4.83^{b}	2.42^{b}	4.72^{b}	2.01^{b}
TF		3.50^{c}	2.48^{b}	3.36^{c}	1.75^{c}
CS	2018	3.22^{a}	1.26^{a}	2.94^{a}	1.05^{a}
S		2.97^{a}	1.41^{b}	2.90^{a}	1.12^{b}
TN		5.35^{b}	2.01^{c}	5.26^{b}	1.98^{c}
TF		4.80^{c}	1.79^{c}	4.56^{c}	1.51^{c}

Variety	Year	Midterm		Harvest	
		tla (m^2)	g_s (mm s^{-1})	tla (m^2)	g_s (mm s^{-1})
CS	2019	2.76^a	1.39^a	2.61^a	1.29^a
S		2.72^a	1.42^a	2.72^a	1.14^a
TN		4.36b	2.05^b	5.00^c	1.75^b
TF		4.28b	2.12^b	4.15^b	1.65^b
	2016	3.67^a	1.69^a	3.68^a	1.38^a
	2017	3.32^b	2.04^b	3.15^b	1.62^b
	2018	4.08^c	1.62^a	3.91^c	1.42^a
	2019	3.78^d	1.75^a	3.51^a	1.37^a

Different superscript letters on the same column and pertaining to varieties grouped by years mean significant difference (Tukey's $HSD_{0.05}$).

Table 5.
Mean separation of total leaf area (tla) and stomatal conductance (g_s) at midterm between veraison and harvest and at harvest across cabernet sauvignon, Syrah, Touriga Nacional, and Touriga Franca from 2016 to 2019.

Variety	Year	°Brix	Titratable acidity (mg L^{-1})	pH	Glucose (g L^{-1})	Fructose (g L^{-1})
CS	2016	25.2^a	3.69^a	3.90^a	113.25^a	84.06^a
S		25.1^a	3.69^a	3.15^a	117.36^a	82.70^a
TN		26.5^b	3.30^b	3.85^a	133.21^b	100.69^b
TF		26.8^b	3.40^b	3.86^a	135.85^b	99.01^b
CS	2017	22.3^a	3.98^a	3.90^a	105.98^a	75.56^a
S		21.6^b	3.94^a	4.20^a	108.79^a	73.68^a
TN		24.5^c	3.89^a	3.85^a	115.60^b	90.15^b
TF		24.2^c	3.87^a	3.77^a	118.44^b	87.75^b
CS	2018	23.6^a	3.87^a	3.88^a	109.02^a	79.90^a
S		23.4^a	3.13^b	4.13^a	105.60^a	77.29^a
TN		24.6^b	3.79^a	3.88^a	130.10^b	92.49^b
TF		25.4^c	3.95^a	3.86^a	127.92^b	89.86^b
CS	2019	23.8^a	3.17^a	3.83^a	106.21^a	80.15^a
S		24.0^a	3.83^a	3.91^a	110.65^a	78.25^a
TN		25.1^b	3.59^a	3.98^a	128.22^b	95.68^b
TF		25.6^b	3.57^a	3.81^a	131.58^b	91.26^b
	2016	25.9^a	3.52^a	3.69^a	124.92^a	91.62^a
	2017	23.2^b	3.92^b	3.93^a	112.20^b	81.79^b
	2018	24.2^c	3.69^c	3.94^a	118.16^b	84.88^b
	2019	25.0^c	3.65^c	3.91^a	121.20^c	85.90^b

Different superscript letters on the same column and pertaining to varieties grouped by years mean significant difference (Tukey's $HSD_{0.05}$).

Table 6.
Mean separation of must characteristics (Brix, acidity. pH and sugars) across cabernet sauvignon, Syrah, Touriga Nacional, and Touriga Franca from 2016 to 2019.

Variety	Year	Malic acid (g L^{-1})	Tartaric acid (g L^{-1})	Total tannins (mg $berry^{-1}$)	Total polyphenols (mg $berry^{-1}$)	Total anthocyanins (mg $berry^{-1}$)
CS	2016	1.20^{a}	4.21^{a}	8.59^{a}	266.0^{a}	2.26^{a}
S		1.23^{a}	4.32^{a}	9.15^{a}	303.0^{a}	2.78^{b}
TN		1.16^{a}	3.65^{b}	5.69^{b}	212.0^{c}	2.02^{c}
TF		1.17^{a}	3.80^{b}	6.81^{b}	237.0^{ac}	2.16^{c}
CS	2017	1.31^{a}	3.63^{a}	6.95^{a}	224.0^{a}	1.89^{a}
S		1.32^{a}	3.70^{a}	7.85^{a}	271.0^{b}	2.47^{b}
TN		1.24^{a}	3.08^{a}	7.01^{a}	238.0^{a}	2.11^{c}
TF		1.26^{a}	3.10^{a}	6.86^{a}	255.0^{ab}	2.55^{b}
CS	2018	2.18^{a}	3.46^{a}	7.40^{a}	240.0^{a}	2.07^{a}
S		2.26^{a}	3.71^{a}	8.22^{b}	275.0^{b}	2.47^{b}
TN		1.85^{a}	2.50^{b}	6.76^{c}	229.0^{c}	2.00^{a}
TF		1.93^{a}	2.31^{b}	6.86^{c}	240.0^{a}	2.18^{a}
CS	2019	1.78^{a}	3.96^{a}	7.99^{a}	251.6^{a}	2.12^{a}
S		2.01^{a}	3.78^{a}	8.54^{a}	278.2^{b}	2.59^{b}
TN		1.98^{a}	2.87^{a}	6.49^{b}	240.1^{a}	2.09^{a}
TF		1.89^{a}	2.56^{a}	6.51^{b}	248.7^{a}	2.11^{a}
	2016	1.19^{a}	3.98^{a}	7.56^{a}	254.5^{a}	2.31^{a}
	2017	1.28^{a}	3.35^{b}	7.17^{b}	247.0^{b}	2.26^{a}
	2018	2.06^{b}	2.98^{c}	7.31^{c}	246.0^{b}	2.18^{b}
	2019	1.31^{a}	2.65^{d}	7.28^{c}	256.4^{a}	2.21^{b}

Different superscript letters on the same column and pertaining to varieties grouped by years mean significant difference (Tukey's $HSD_{0.05}$).

Table 7.
Mean separation of must characteristics (organic acids, tannins and phenols) across cabernet sauvignon, Syrah, Touriga Nacional, and Touriga Franca from 2016 to 2019.

4. Discussion

The trend for higher temperatures from 2003 to 2019 was already observed in the period 1980 to 2009 in the DDR and its correlation with earlier phenological events was statistically significant [15] as it was also reported by other authors [10, 21, 44]. Shifts in phenology are a clear biological signal of climate change [45]. Although the observed pattern of precipitation from 2003 to 2019 is not aligned with the predictions for the Mediterranean, the tendency for higher rainfall intensity during winter might aggravate the drought stress during summer because infiltration and soil water storage might be negatively affected. Drought associated to elevated temperatures can cause photoinhibition, increase the reduction of growth, yield and alter the berry composition [20, 46].

Budbreaking occurred 71 days after 1 January, with no significant difference among the four varieties, and it was very similar with the occurrence between 2003 to 2019. Flowering was shifting to earlier dates from the first days of June to late May that is consistent with the observation in DDR and other wine regions of the world

[15, 47, 48], however, the two exotic varieties CS and S flowered about 10 days earlier than TN and TF and the late ones need higher temperature accumulation (GDD) to start flowering. The date for the posterior phenological event was set by the flowering date as the time span for reaching the next stage was the same for every variety, either exotic or autochthonous, but always later for TN and TF and the associated GDD was necessarily larger because the average temperatures were progressively higher till earlier August. Climate alone does not explain all phenological variation as different grape cultivars grown under similar conditions still show large variations in the timing of different events [21]. The number of days and the GDD necessary to complete the growing cycle of these four varieties were close to the figures reported by other authors but TN and TF can grow over a wider range of thermal conditions that favors their adaptability to different climatic conditions [29, 49–51].

Commercial practice on winter pruning sets the number of potential developed inflorescences to an equal value, independently of grape variety, but the number of flowers per inflorescence is determined by genetics and environmental factors [52]. In this experiment all plants were subjected to the same environment, thus it is reasonable to assume that the lower number of flowers per inflorescence in SC and S than in TN and TF is a varietal characteristic and it has a large influence on fruit setting [47]. Fruit set also increases with larger leaf area [53], a phenomenon recorded, that together with cluster number accounts for about 90% of grapevine yield variation [54]. The higher yields obtained by TN and TF over SC and F are consistent with these facts. However, the positive influence of higher temperatures on fruit set [53] was not observed as the number of flowers per inflorescence and the yields were lower in 2017 the hottest year; other factors are at play that could not be determined in this experiment.

The higher productivity of TN and TF were supported by their larger total leaf area and superior stomatal conductance. As the envelope volume of the canopies were limited by mechanical trimming, tla was related to the number of leaves, eventually their size (not measured), which was larger for TN and TF. Low g_s reduce the amount of CO_2 available impairing the photosynthetic rate [55] and together with smaller leaf area probably results in a lower photosynthetic capacity for SC and S comparatively to TN and TF which can partially explain better yields for the regional varieties. The mechanistic adaptation of grapevines varieties to summer stress is mediated by the stomatal sensitivity and the anisohydric cultivars, like TN, are better adapted than iso- or near-isohydric ones, like Syrah, because they can maintain high values of g_s regardless of their water status [13, 14]. However, there is no consensus among innumerous authors about the classification of each variety regarding their water stress behavior and the same variety can have both iso- and anisohydric stomatal responses to water deficits [56].

Temperature displays various effects on the physiology of the grapevine fruit. High temperatures accelerate organic acids breakdown [57], a decrease in anthocyanins with possible variations in acylation in red-berry cultivars [58] that are associated with changes in aromatic potential [12]. Moderate warming favors soluble solids concentration (mostly primary sugars) [59] whereas very high temperatures can impede sugar accumulation [57].

All varieties show a similar effect of organic acids degradation by elevated temperatures as a far below the value of titratable acidity from the preferable 6 to 7 mg L^{-1} equivalent of tartaric acid for producing high-end wines [60]. The lack of must acidity has to be corrected in the winery by the addition of tartaric acid, a common commercial practice.

Soluble solids concentration is high, about 21 to 27°Brix [61], given the concentration of primary sugars, glucose, and fructose, as it is expected from a warm climate.

TN and TF accumulated more primary sugars than SC and S resulting in higher °Brix. From the financial perspective of the growers, higher °Brix is a clear advantage because the berry prices vary proportionally to sugar content; in addition, TN and TF have better yields making them a more attractive crop. However, SC and S retain a better organoleptic profile as their tannins and polyphenols reach higher concentrations at harvest. Under prevailing conditions during the duration of the experiment, the temperature threshold for reduced accumulation of soluble solids was not yet reached as the Brix at harvest is still high.

Must pH display values commonly found with no significant differences among the four varieties and a relationship between pH and genetic makeup of the grapevine was never established [61]. The anthocyanins concentration showed no clear distribution among the cultivars although their actual values might reflect the reduced capacity for their synthesis under the observed temperatures [61] but other authors report that temperature does not modify anthocyanins concentration [20].

The results suggest clearly that the effect of climate on grape yield and quality is cultivar-dependent as reported [20]. Some authors mention that most of the differences in the composition of individual fruit at harvest can be related to differences in the date of flowering [47], a phenomenon observed here when local varieties flowered about 10 days earlier than exotic ones, thus later berries would develop under different weather conditions.

Autochthonous varieties might have high adaptability because they are grown over a large range of thermal conditions [29] providing adequate water supply [62]. Thus, rainfed viticulture might not be possible in a near future given that the actual conditions during the Mediterranean summer that are already severe water-stressed [63] and increased stress is forecasted.

5. Conclusions

Observed weather conditions over several years coupled with reports by other authors clearly indicate a climate shift towards a more stressful environment in the DDR as annual temperature increased 2°C in 16 years. However, total annual precipitation increased contrary to the general predictions but it was due to higher rainfalls only in winter that favors runoff at expense of infiltration. The phenological advance of grapevines was imparted mostly by the advance of about 10 to 15 days in flowering and it is an unmistakable biological signal of climate change which has also altered the composition of the fruits. Autochthonous varieties have rendered results that suggest a better adaptation to temperature stress-producing higher yields, about 2000 kg ha^{-1} higher, and concentrations of sugars that are economically important, about 1.5 to 2°Brix higher, than exotic cultivars. However, the exotic varieties, so far, have kept a better organoleptic profile with higher concentrations in berries of malic and tartaric acids, total tannins, and total polyphenols. Given the predicted increase in stressful conditions in a near future, new plantations with autochthonous varieties will be more likely to withstand coming shifting conditions but, probably, only with irrigation.

Acknowledgements

This work was possible with the assistance of OLAZABAL and SONS – Quinta do Vale Meão.

DOI: http://dx.doi.org/10.5772/intechopen.101866

Declaration of interests

The authors declare no conflict of interests or any undue benefit from the present work.

Author details

Manuel T. Oliveira* and Ana A. Oliveira
UTAD, Vila Real, Portugal

*Address all correspondence to: mto@utad.pt

References

[1] Bonada M, Sadras VO. Review: Critical appraisal of methods to investigate the effect of temperature on grapevine berry composition. Australian Journal of Grape and Wine Research. 2015;**21**(1):1-17

[2] Mosedale JR et al. Climate change impacts and adaptive strategies: Lessons from the grapevine. Global Change Biology. 2016;**22**(11):3814-3828

[3] Jacob D et al. Regional climate downscaling over Europe: perspectives from the EURO-CORDEX community. Regional Environmental Change. 2020;**20**(2):51

[4] JRC. Worrying Effects of Accelerating Climate Change on the Mediterranean Basin. 2018. [cited 2019 September]; Available from: https://ec.europa.eu/jrc/en/science-update/worrying-effects-accelerating-climate-change-mediterranean-basin

[5] Ramos MC, Jones GV, Martínez-Casasnovas JA. Structure and trends in climate parameters affecting winegrape production in northeast Spain. Climate Research. 2008;**38**:1-15

[6] Gallego MC et al. Trends in frequency indices of daily precipitation over the Iberian Peninsula during the last century. Journal of Geophysical Research. 2011;**116**(D2):1-18

[7] Hannah L et al. Climate change, wine, and conservation. Proceedings of the National Academy of Sciences of the United States of America. 2013;**110**(17):6907-6912

[8] Santos M et al. Recent and future changes of precipitation extremes in mainland Portugal. Theoretical and Applied Climatology. 2018;**137**(1-2):1305-1319

[9] Greer DH, Weedon MM. High temperatures affect gas exchange, growth and ripening processes of 'Semillon' vines. Acta Horticulturae. 2017;**1157**:183-190

[10] Fila G et al. A comparison of different modelling solutions for studying grapevine phenology under present and future climate scenarios. Agricultural and Forest Meteorology. 2014;**195-196**:192-205

[11] Ramos MC. Projection of phenology response to climate change in rainfed vineyards in north-east Spain. Agricultural and Forest Meteorology. 2017;**247**:104-115

[12] Mira de Orduña R. Climate change associated effects on grape and wine quality and production. Food Research International. 2010;**43**(7):1844-1855

[13] Palliotti A et al. Changes in vineyard establishment and canopy management urged by earlier climate-related grape ripening: A review. Scientia Horticulturae. 2014;**178**:43-54

[14] Chaves MM et al. Grapevine under deficit irrigation: Hints from physiological and molecular data. Annals of Botany. 2010;**105**(5):661-676

[15] Real AC et al. Partitioning the grapevine growing season in the Douro Valley of Portugal: Accumulated heat better than calendar dates. International Journal of Biometeorology. 2015;**59**(8):1045-1059

[16] Oliveira M. Viticulture in warmer climates: Mitigating environmental stress in Douro Region, Portugal. In: Grapes and Wines - Advances in Production, Processing, Analysis and Valorization. London: IntechOpen; 2018

[17] Oliveira MT, Sousa TA. Organic acids and sugars in musts of irrigated grapevines in Northeast Portugal. Journal of Wine Research. 2009;**20**(1):1-13

[18] Oliveira MT, Freitas VD, Sousa TA. Water use efficiency and must quality of irrigated grapevines of north-eastern Portugal. Archives of Agronomy and Soil Science. 2012;**58**(8):871-886

[19] Vilanova M et al. Assessment fertigation effects on chemical composition of Vitis vinifera L. cv. Albarino. Food Chemistry. 2019;**278**:636-643

[20] Kizildeniz T et al. Effects of climate change including elevated CO2 concentration, temperature and water deficit on growth, water status, and yield quality of grapevine (Vitis vinifera L.) cultivars. Agricultural Water Management. 2015;**159**:155-164

[21] Wolkovich EM et al. Phenological diversity provides opportunities for climate change adaptation in winegrapes. Journal of Ecology. 2017;**105**(4):905-912

[22] Wang Z et al. Effects of different statistical distribution and threshold criteria in extreme precipitation modelling over global land areas. International Journal of Climatology. 2019;**40**(3):1838-1850

[23] Mozell MR, Thach L. The impact of climate change on the global wine industry: Challenges & solutions. Wine Economics and Policy. 2014;**3**(2):81-89

[24] Cunha J et al. A identidade das castas de videira portuguesas aptas à produção de vinho no contexto ibérico e europeu. Viticultura. 2017;**Jan-Fev-Mar**:18-25

[25] Costa P. Castas Portuguesas. Um Património Único. 2012 [cited 2019 September 2019]. Available from: http://www.advid.pt/imagens/comunicacoes/14171928355324.pdf

[26] Costa E et al. Influence of wine region provenance on phenolic composition, antioxidant capacity and radical scavenger activity of traditional Portuguese red grape varieties. European Food Research and Technology. 2015;**241**(1):61-73

[27] Augusto D, Oliveira AA, Falco V. Uncovering Northeast Portugal grapevine's varietal legacy. Vitis. 2019;**58**(Special Issue):89-93

[28] Prata-Sena M et al. The terroir of Port wine: Two hundred and sixty years of history. Food Chemistry. 2018;**257**:388-398

[29] Fraga H et al. Climatic suitability of Portuguese grapevine varieties and climate change adaptation. International Journal of Climatology. 2016;**36**(1):1-12

[30] De Pinho PG et al. Further insights into the floral character of Touriga Nacional wines. Journal of Food Science. 2007;**72**(6):S396-S401

[31] Guerra J and Abade E. Castas do Douro. Caracterização Enológica de Castas Tintas. 2008 [cited 2019 November]. Available from: http://www.drapn.min-agricultura.pt/drapn/conteudos/fil_trab/Castas%20Tintas%20do%20Douro.pdf

[32] OIV. Vine Varieties Distribution In The World. 2017. Available from: http://www.oiv.int/public/medias/5336/infographie-focus-oiv-2017-new.pdf

[33] Jarvis C et al. Relationship between viticultural climatic indices and grape maturity in Australia. International Journal of Biometeorology. 2017;**61**(10):1849-1862

[34] Fraga H et al. What is the impact of heatwaves on European viticulture? A modelling assessment. Applied Sciences. 2020;**10**(9):3030

[35] Soil Survey Staff. 2014 Keys to Soil Taxonomy. N.R.C.S. United States. Washington: Department of Agriculture; 2014. p. 360

[36] Coombe BG. Growth stages of the grapevine: Adoption of a system for identifying grapevine growth stages. Australian Journal of Grape and Wine Research. 1995;**1**(2):104-110

[37] Aquino A et al. vitisFlower(R): Development and testing of a novel android-smartphone application for assessing the number of grapevine flowers per inflorescence using artificial vision techniques. Sensors (Basel). 2015;**15**(9):21204-21218

[38] Oliveira M, Santos M. A semi-empirical method to estimate canopy leaf area of vineyards. American Journal of Enology and Viticulture. 1995;**46**(3):389-391

[39] OIV. International Oenological Codex. Paris: OIV; 2021. p. 825

[40] Gu S. Growing degree hours - a simple, accurate, and precise protocol to approximate growing heat summation for grapevines. International Journal of Biometeorology. 2016;**60**(8):1123-1134

[41] IPCC. Global Warming of 1.5 °C. 2019 [cited 2019 November]. Available from: https://www.ipcc.ch/sr15/

[42] Jones GV, Alves F. Impact of climate change on wine production: a global overview and regional assessment in the Douro Valley of Portugal. International Journal of Global Warming. 2012;**4**(3/4):383-406

[43] Deitch M, Sapundjieff M, Feirer S. Characterizing precipitation variability and trends in the World's Mediterranean-Climate areas. Water. 2017;**9**(4):259

[44] Bock A et al. Changes in the phenology and composition of wine from Franconia, Germany. Climate Research. 2011;**50**(1):69-81

[45] Sadras VO, Moran MA. Nonlinear effects of elevated temperature on grapevine phenology. Agricultural and Forest Meteorology. 2013;**173**:107-115

[46] Cramer GR. Abiotic stress and plant responses from the whole vine to the genes. Australian Journal of Grape and Wine Research. 2010;**16**:86-93

[47] Eltom M et al. Pre-budburst temperature influences the inner and outer arm morphology, phenology, flower number, fruitset, TSS accumulation and variability of Vitis vinifera L. Sauvignon Blanc bunches. Australian Journal of Grape and Wine Research. 2017;**23**(2):280-286

[48] Costa R et al. Grapevine phenology of cv. Touriga Franca and Touriga Nacional in the Douro Wine Region: Modelling and Climate Change Projections. Agronomy. 2019;**9**(4):210

[49] Ramos MC, Jones GV, Yuste J. Phenology of Tempranillo and Cabernet-Sauvignon varieties cultivated in the Ribera del Duero DO: Observed variability and predictions under climate change scenarios. OENO One. 2018;**52**(1):2119

[50] Miele A. Rootstock-scion interaction:6. Phenology, chilling and heat requirements of Cabernet Sauvignon grapevine. Revista Brasileira de Fruticultura. 2019;**41**(6):1-14

[51] Sartor S et al. Particularities of Syrah wines from different growing regions of Southern Brazil: Grapevine phenology and bioactive compounds. Journal of Food Science and Technology. 2017;**54**(6):1414-1424

[52] Gourieroux AM et al. Vascular development of the grapevine (Vitis vinifera L.) inflorescence rachis in response to flower number, plant growth regulators and defoliation. Journal of Plant Research. 2017;**130**(5):873-883

[53] Keller M, Tarara JM, Mills LJ. Spring temperatures alter reproductive development in grapevines. Australian Journal of Grape and Wine Research. 2010;**16**(3):445-454

[54] Guilpart N, Metay A, Gary C. Grapevine bud fertility and number of berries per bunch are determined by water and nitrogen stress around flowering in the previous year. European Journal of Agronomy. 2014;**54**:9-20

[55] Zamorano D et al. Improved physiological performance in grapevine (Vitis vinifera L.) cv. Cabernet Sauvignon facing recurrent drought stress. Australian Journal of Grape and Wine Research. 2021;**27**(3):258-268

[56] Levin AD, Williams LE, Matthews MA. A continuum of stomatal responses to water deficits among 17 wine grape cultivars (Vitis vinifera). Functional Plant Biology. 2019;**47**(1):11-25

[57] Gashu K et al. Temperature Shift Between Vineyards Modulates Berry Phenology and Primary Metabolism in a Varietal Collection of Wine Grapevine. Frontiers in Plant Science. 2020;**11**:588739

[58] Greer DH, Weston C. Heat stress affects flowering, berry growth, sugar accumulation and photosynthesis of Vitis vinifera cv. Semillon grapevines grown in a controlled environment. Functional Plant Biology. 2010;**37**(3):206-2014

[59] Keller M. Managing grapevines to optimise fruit development in a challenging environment: A climate change primer for viticulturists. Australian Journal of Grape and Wine Research. 2010;**16**:56-69

[60] Esteban MA, Villanueva MJ, Lissarrage JR. Effect of irrigation on changes in berry composition of tempranillo during maturation. Sugars, Organic Acids, and Mineral Elements. American Journal of Enology and Viticulture. 1999;**50**(4):418-434

[61] Gomes E, Maillot P, Duchene E. Molecular tools for adapting viticulture to climate change. Frontiers in Plant Science. 2021;**12**:633846

[62] Carvalho LC et al. Differential physiological response of the grapevine varieties Touriga Nacional and Trincadeira to combined heat, drought and light stresses. Plant Biology (Stuttgart, Germany). 2016;**18**(Suppl. 1): 101-111

[63] Costa JM et al. Modern viticulture in southern Europe: Vulnerabilities and strategies for adaptation to water scarcity. Agricultural Water Management. 2016;**164**:5-18

Chapter 5

Climate Change and Food System in Kenya: Challenges and Opportunities

Festus Kelonye and Godfrey Juma

Abstract

Climate change is long-term statistical change in climate variables whose impacts are evident on value chains and food systems in general. This book chapter examines the challenges in the Kenyan food system associated with increasing climate variability and factors that can be considered to achieve agro-food systems transformation for sustainability. The study observed the need for digital infrastructure enhancement; development of data sharing platforms; accelerated climate change mitigation and adaption actions; rural infrastructural development, fragmentation of value chains and farmers' societies; value addition to increase shelf life of agricultural products; universal crop and livestock insurance; promotion of agroecological farming practices, including agroforestry; involvement of youths in agricultural practices through government-supported initiatives; systematic improvement of policies toward a stronger policy framework to regulate imports and maximize exports; empowering populations on the need to consume local foods; engagement in consumer protection initiatives as important for enhanced food production in a changing environment.

Keywords: agro-food systems; climate change, challenges, opportunities

1. Introduction

Climate change is an evolving concept with future significance across all sectors [1]. The vulnerable sectors include but are not limited to the following viz.; transport, agriculture, water, health, education, etc. as examined by [2–6], respectively. According to the food and agriculture organization [7], Food systems (FS) encompass the entire range of actors and their interlinked value-adding activities involved in the production, aggregation, processing, distribution, consumption, and disposal of food products that originate from agriculture, forestry or fisheries, and parts of the broader economic, societal and natural environments in which they are embedded. In this regard, food systems are characterized by drivers that are environmental or biophysical and/or socioeconomic in nature, anchored on the three pillars of sustainable development. The food systems should have the ability to address the four dimensions of food security namely availability, access, stability, and utilization [8].

The need for proper food systems is in view of the increasing demand for sustainable healthy diets across the globe amidst the increasing population and consequently, the demand for healthy, accessible, and sustainable food. This food is part of a food system that can be affected by changes in biodiversity, climate, and many other global phenomena. This calls for the continuous need to establish evidence-based solutions to achieve resilient food systems through approaches, such as ensuring that there is enough food for everyone, there is access to a healthy diet, a fair food system, and a future-proof planet [9].

Challenges in the food system that calls for the need for food systems transformation have been studied globally. Studies by Dinesh et al. [10] implemented a theory of change geared toward transforming food systems under a changing climate. The study identified focal actions for food systems transformation that include empowering women, youths, consumer and farmer's organizations; acceleration of digitally facilitated climate services; adoption of climate-smart agriculture and technology; financial innovativeness in agricultural investments; reshaping supply chains, policies, and institutions; defragmentation in knowledge and food security.

Cascading these findings into implementable frameworks still faces challenges in the wake of different perspectives of governance among countries and weak policies and institutions [11]. In general, actions to transform food systems have been reported and summarized in **Figure 1**.

The systems approach is a current debate in food systems transformation where all food system drivers are considered in every stage of agricultural food systems [13]. Decisions and practices that maximize synergies among the drivers and minimize trade-offs are better suited for sustainable agricultural production [14].

The insurgency of pandemics, including COVID-19 further raises the need for resilient, sustainable, and inclusive food systems to realize sustainable development goals. *The food system entails all the processes through which food goes from* production to consumption and is composed of inputs, production, processing, retail, marketing, and consumption. The food system also includes external factors that influence them, such as policies, environment, economic and sociocultural factors. It is characterized by a system approach where all relevant factors and actors are considered in all activities that guide food production, access, stability, and utilization.

There are many reasons why we need to change the current food systems. One out of every three people suffers from malnutrition, at least 794 million people suffer from hunger, 2 billion people do not have sufficient access to vitamins and minerals for sufficient growth and development while 1.9 billion people overeat of which 600 million of those people are obese. As a result, more people suffer from conditions, such as type 2 diabetes associated with nutritional conditions [15]. Billions of people consume food too rich in fat, sugar, salt, and meat that impacts on health and the environment, for example, by being a risk factor to heart deceases and high greenhouse emissions emanating from meat production. Lack of dietary diversity has also been observed in populations where 75% of food consumed across the globe comes from only 12 plants, including rice, corn, and wheat; and from five animal species, including cattle, chicken, and pigs. Studies have also observed that one-third of food produced across the globe is wasted even as pressure on natural resources is increasing including dryness and pollution of sources of fresh water, 33% of soils are degraded, biodiversity is threatened as tropical forests are disappearing. This pressure is intensified by increasing climate variability and/or change. These limitations point out the need to transform the food system so that each step of the food system, including production, processing, distribution, consumption, should be adjusted to ensure

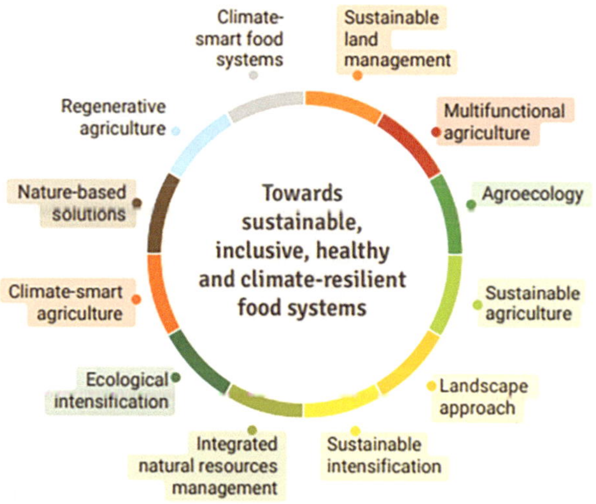

Figure 1.
Actions to transform food systems [12].

healthier food to growing populations while reducing the environmental impact and food wastage [16]. Empowering people to eat local food thereby supporting the food growers in each country is a challenge that will cut off so many habits and lifestyles from many people.

Developing countries, inability to add value to what has been produced to increase the shelf life of foods and make them available at off-seasons is one of the challenges alongside infrastructural issues which is the responsibility of governments and needs a collective responsibility to solve them.

Africa is vulnerable to the impacts of climate change associated with multiple biophysical, socioeconomic, and political stresses that interact to reduce the region's adaptive capacity to the impacts of climate change mediated environmental hazard risks [17]. Climate change can significantly impact both rural and urban food systems by being attributed to conditions that spoil food, disrupt food production and transport processes, and therefore contribute to high market prices for food among populations.

Kenya is composed of a population of 57.57 million people [18], an increase from 40.9 million people in the year 2009. Out of this population, at least **14,461,521 people live in urban centers representing a** 4.1% increase from 2018. In a population of at least 7 million people under 5 years; 1.8 million have been identified with cases of malnutrition with an increasing decease burden of diabetes mellitus among Kenyans. Increasing cases of drought-induced food insecurity have been documented with at least 2,147,889 Kenyans in need of immediate food assistance in the year 2021 [19]; even as increasing cases of obesity have been observed with a prevalence rate of 60.3% among urban residents and 19.5% among rural residents. An estimated 99 kilograms of food is discarded by every Kenyan annually with an annual country-wide food wastage of 5.2 million tonnes [20]. Climate change is equally evident in Kenya with widespread impacts that include enhanced drought and flood events [21]. Consequently, there are limitations of the Kenyan food system that point out the need for transformation. **Figure 2** shows the interrelationship between various components of a food system and how they interact with each other.

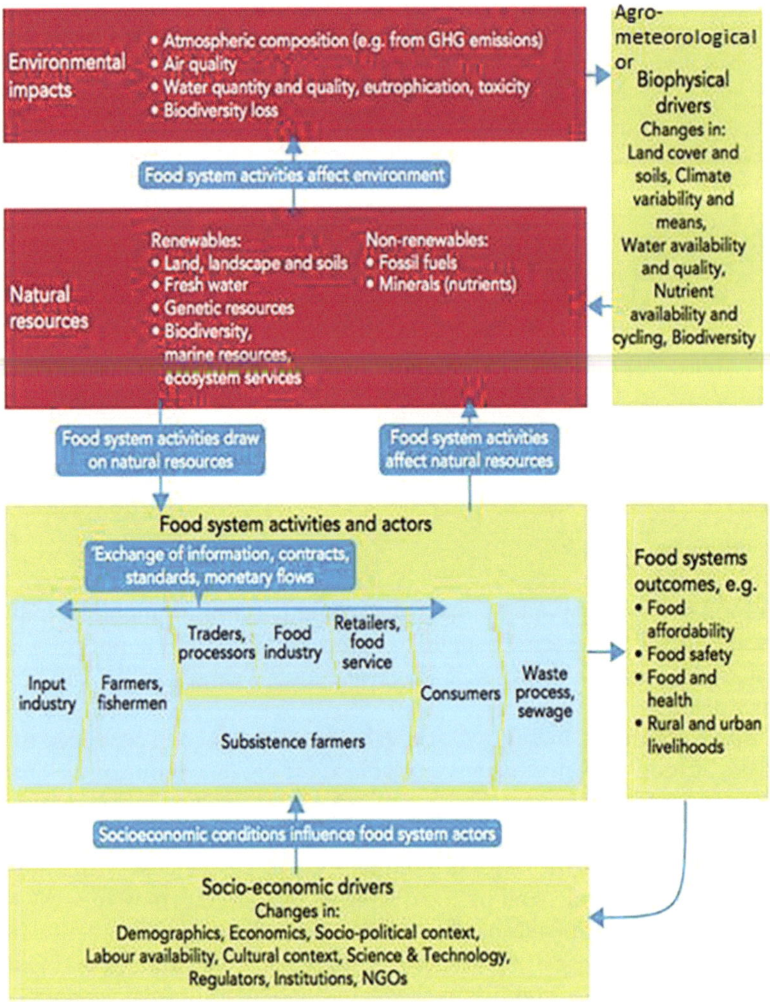

Figure 2.
Relationships between components of a food system.

2. Enhancing food systems in Kenya

The food system approach can help in integrating all actors and players in the government of Kenya on a common platform that enables all stakeholders involved to have or benefit from sustainable production systems. This will not only benefit the current generations but the future as we save our planet from the adverse effects of climatic shocks. Poor soil management, such as excessive use of inorganic fertilizer, leads to soil degradation, thus affecting the quality and quantity of food produced. This, in turn, affects the availability of food and also leads to socioeconomic distress where small-scale farmers are not able to break even.

The Kenyan Food system is married by challenges that include rural to urban migration of energetic and potentially food-producing population, high poverty levels among smallholder food producers that may not afford firm inputs, socioeconomic inequalities, cheap food imports at the expense of local production, low-quality seeds sneaking in their way to farmers and increasing impacts of climate variability and/or change.

According to a food systems analysis by Wagengen Centre for Development, the following key themes identified in Nigeria can be applied in Kenya to achieve resilient food systems namely transforming the agricultural sector; employment of youth and women; access to finance; climate change adaptation and mitigation; agribusiness, value chain development, and logistics. **Figure 3** shows a Food Systems Decision Support Tool (FSDS) developed by KIT and Wageningen Economic Research to provide a scan of the food system and identify leverage points to inform policy recommendations for the Dutch Government. The tool consists of seven steps namely defining the policy objectives, mapping the agro-food system relevant to these policy objectives, identifying the causal processes underlying the agrifood system, determining archetypes in system behavior of the agro-food system, identifying actionable leverage points within the agro-food system, defining relevant actors and their influence and interest to address leverage points, based on leverage points, policy objectives and relevant actors and provide policy recommendation.

We use the FSDS to determine the SDG-guided objectives for transforming the Kenyan food systems and determine the indicators, trade-offs, and synergies for their sustainability.

The food system in Kenya is identifiable with different drivers that are interrelated, as shown in **Figure 4**. Institutionalization is characterized by the development of institutional infrastructure and mandates that enhance the strengthening of legal and policy frameworks, including their implementation for sustainable development. The process of institutionalization in Kenya is married with challenges that include lack of policies for universal agricultural funding; lack of unitary or stronger farmer organizations, weak policies for protecting consumers and markets, and lack of universal crop and livestock insurance.

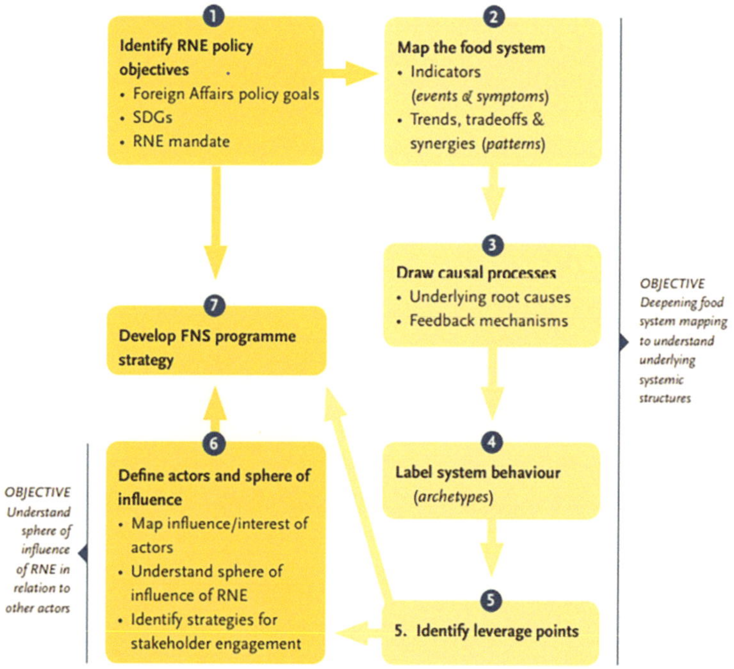

Figure 3.
Food systems decision support tool.

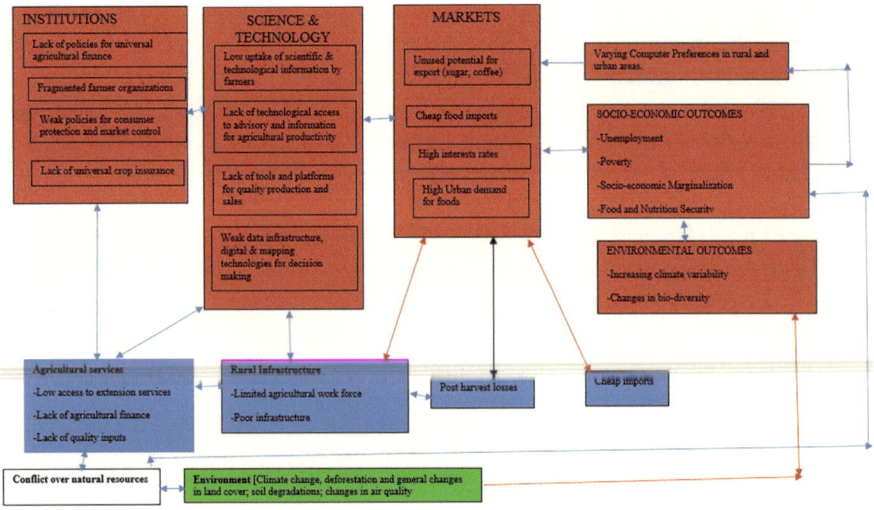

Figure 4.
Interrelationship between different drivers and outcomes of a Kenyan food system.

The uptake of scientific and technological information has remained relatively low compared to developed countries with inadequate access to agricultural advisory and information for enhanced agricultural productivity alongside lack of tools and even platforms that accelerate productivity and sales of farm produce coupled with weak digital and mapping infrastructure for agricultural decision-making.

High-interest rates and cheap imports have jeopardized agricultural productivity in Kenya even as urban demand for food is continuously increasing while young people of potential productivity migrate from rural to urban areas limiting agricultural production in rural areas. Further, the agricultural productivity in Kenya is greatly compromised by poor rural infrastructure [22].

Economic inequality has characteristically led to food consumption preferences among the urban and rural populations where rural populations have preferred traditional food supplies while urban ones manufactured food whose utilization has been poor in dietary diversity and precursors for malnutrition and obesity [15].

Increasing climate change as an environmental driver further impacts agricultural productivity through an erratic climate that compromises the sustainability of crops and livestock reducing food production and leading to food insecurity [23, 24]. The environmentally mediated socioeconomic impacts further increase the overall vulnerability of the agricultural communities to disaster risks associated with the environment and other drivers [25]. There is, therefore, the need for deliberate efforts to integrate climate change mitigation with agricultural practices alongside climate change adaptation activities for enhanced food production in Kenya.

3. Conclusions

The study examined challenges that can be addressed to enhance the Kenyan agricultural productivity that includes fragmented market and value chains and small regionally separated farmer groups and organizations; slow digital technology uptake; infrastructural challenges; science and technology challenges; market changes that can be associated with various habits, such as cravings for imported food and outcomes that include socioeconomic and environmental factors. The environmental factors include increasing climate variability and/or climate change.

The study observed that there is a need to defragment value and market chains and farmer organizations; enhance digital infrastructure and data sharing platforms for agricultural marketing and decision-making; accelerate rural infrastructural development, practice value addition of agricultural products to increase their quality and shelf life; involve youths in agricultural practices and decisions; Involve in agroecological practices; provide universal crop and livestock insurance; develop a stronger policy framework to regulate imports and maximize exports; empower Kenyans on the need to consume local foods; engage in initiatives for consumer protection and upscale climate change mitigation and adaptation actions.

Author details

Festus Kelonye[1*] and Godfrey Juma[2]

1 Department of Biological and Environmental Sciences, Kibabii University, Bungoma, Kenya

2 Department of Mathematics, Kibabii University, Bungoma, Kenya

*Address all correspondence to: fkelonye@yahoo.com

References

[1] Kundzewicz ZW, Piniewski M, Mezghani A, Okruszko T, Pińskwar I, Kardel I, et al. Assessment of climate change and associated impact on selected sectors in Poland. Acta Geophysica. 2018;**66**(6):1509-1523

[2] Creutzig F, Jochem P, Edelenbosch OY, Mattauch L, van Vuuren DP, McCollum D, et al. Transport: A roadblock to climate change mitigation? Science. 2015;**350**(6263): 911-912

[3] Clapp J, Newell P, Brent ZW. The global political economy of climate change, agriculture and food systems. The Journal of Peasant Studies. 2018;**45**(1):80-88

[4] Singh C, Bazaz A, Ley D, Ford J, Revi A. Assessing the feasibility of climate change adaptation options in the water sector: Examples from rural and urban landscapes. Water Security. 2020;**11**:100071

[5] Otieno OD, Godfrey JS. Preparedness and response to the health impacts of climate change in Kenya. BEST: International Journal of Humanities, Arts, Medicine and Sciences (BEST: IJHAMS). 2020;*8*:9-20

[6] Atuahene S, Kong Y, Bentum-Micah G. COVID-19 pandemic, economic loses and education sector management. Quantitative Economics and Management Studies. 2020;**1**(2):103-109

[7] Food and Agriculture Organization (FAO). Sustainable Food Systems Report. 2018. Available from: Sustainable food systems: Concept and framework (fao.org) [Accessed: November 24, 2021]

[8] Guiné RPF, de Jesus Pato ML, da Costa CA, Aguiar da Costa DVT, da Silva PBC, Martinho VJPD. Food security and sustainability: Discussing the four pillars to encompass other dimensions. Food. 2021;**10**(11):2-26

[9] Wagengen University of Research (WUR) Food Systems Decision Support Tool. 2020

[10] Dinesh D, Hegger DL, Klerkx L, Vervoort J, Campbell BM, Driessen PP. Enacting theories of change for food systems transformation under climate change. Global Food Security. 2021;**1**(31):100583

[11] Holt-Giménez E. Food security, food justice, or food sovereignty. In: Cultivating Food Justice. Race, Class, and Sustainability. Cambridge, MA: MIT Press; 2011. pp. 309-330

[12] Steiner A, Aguilar G, Bomba K, Bonilla JP, Campbell A, Echeverria R, et al. Actions to Transform Food Systems under Climate Change. Wageningen, The Netherlands: CGIAR Research Program on Climate Change, Agriculture and Food Security (CCAFS); 2020

[13] Gill M, Den Boer AC, Kok KP, Breda J, Cahill J, Callenius C, Caron P, Damianova Z, Gurinovic MA, Lähteenmäki L, Lang T, et al. A systems approach to research and innovation for food systems transformation. FIT4FOOD2030. 2018. Available from: https://fit4food2030.eu/reports-publications/#PolicyBriefs

[14] Jagustović R, Papachristos G, Zougmoré RB, Kotir JH, Kessler A, Ouédraogo M, et al. Better before worse trajectories in food systems? An investigation of synergies and trade-offs

through climate-smart agriculture and system dynamics. Agricultural Systems. 2021;**1**(190):103131

[15] Wan S. Study on food nutrition and association with diabetes mellitus. In: IOP Conference Series: Earth and Environmental Science. Vol. 615. IOP Publishing; 2020. p. 012109

[16] Godfray HCJ, Crute IR, Haddad L, Lawrence D, Muir JF, Nisbett N, et al. The future of the global food system. Philosophical Transactions of the Royal Society B. 2010;**365**(1554):2769-2777

[17] Connolly-Boutin L, Smit B. Climate change, food security, and livelihoods in sub-Saharan Africa. Regional Environmental Change. 2016;**16**(2):385-399

[18] Trizer M. Kenya population and housing census results. Kenya National Bureau of Statistics; 2019

[19] National Drought Management Authority (NDMA). Short Rains Assessment Report. 2021

[20] The Food Waste Index Report 2021 by UNEP and Waste and Resources Action Programme (WRAP)

[21] Kogo BK, Kumar L, Koech R. Climate change and variability in Kenya: A review of impacts on agriculture and food security. Environment, Development and Sustainability. 2021; **23**(1):1-21

[22] Renkow M, Hallstrom DG, Karanja DD. Rural infrastructure, transactions costs and market participation in Kenya. Journal of Development Economics. 2004;**73**(1):349-367

[23] Ayah R, Joshi MD, Wanjiru R, Njau EK, Otieno CF, Njeru EK, et al. A population-based survey of prevalence of diabetes and correlates in an urban slum community in Nairobi, Kenya. BMC Public Health. 2013;**13**(1):1-11

[24] Kabubo-Mariara J, Kabara M. Climate change and food security in Kenya. In: Berck CS, Berck P, Falco SD, editors. Agricultural Adaptation to Climate Change in Africa. Routledge: RFF Press; 2018. pp. 55-80

[25] OTIENO OD, GODFREY JS. Preparedness and response to the health impacts of climate change in Kenya. BEST: International Journal of Humanities, Arts, Medicine and Sciences (BEST). 2020;**8**:9-20

Section 2

Ecosystem

Chapter 6

Spatio-Temporal Variation of Ecosystem Services and Its Trade-off Relationships in Southwest Guangxi

Yichao Tian and Yongwei Yang

Abstract

Identifying the mutual relationship between ecosystem services in southwest Guangxi can jointly optimize a variety of services to avoid damaging others while improving one service, which is of great significance for promoting the sustainable management of regional ecosystem, guiding the rational development of natural resources and improving human well-being. Based on remote sensing data, land use data, meteorological data and DEM data, with the support of Carnegie-Ames-Stanford Approach (CASA) model, Integrated Valuation of Ecosystem Services and Trade-offs (InVEST) model and Revised Universal Soil Loss Equation (RUSLE) model, this paper studies the changing characteristics of typical ecosystem services in southwest Guangxi and explores the mutual relationship between different ecosystem services. The results showed that the mean change trend of the whole vegetation net primary productivity (NPP) has been increasing in the study area over the past 19 years. In the past 19 years, water conservation in southwest Guangxi has shown a fluctuating upward trend, with the growth rate of water conservation quality 255.88 mm/hm^{-2}·a^{-1}. During the study period, the range of soil retention variation to the total of 65.38–96.88 t/hm^{-2}·a^{-1} increased 22.73 t/hm^{-2}·a^{-1}, with a mean of 79.19 t/hm^{-2}·a^{-1}. Vegetation NPP in the study area is synergistic with soil conservation and water conservation, and soil conservation with water conservation as well.

Keywords: ecosystem services, net primary productivity, water conservation, soil conservation, trade-off relationships

1. Introduction

Ecosystem services refer to the benefits that human beings can directly or indirectly obtain from ecosystems according to their own development needs, which mainly transfer energy and materials to social and economic systems, accept and transform waste in social and economic systems, as well as the direct services that ecosystems provide food, biomass, clean air, water and other resources [1]. Ecosystem as the basic unit of the biosphere, including maintaining human survival and development of food, freshwater, production and living raw materials supply services, also contains maintaining the dynamic balance of ecosystem and environment primary productivity, soil formation, climate regulation support services and regulation services, as well as pleasant human spiritual and cultural level of aesthetic, entertainment, tourism and other cultural services [2–4]. Trade-off refers to the situation that the supply of some ecosystem services is reduced by the increased consumption of other types of ecosystem services, showing a trend of one after another [5]. Ecosystem has a great impact on people's lives, and there are certain links between various ecosystem services [6]. Therefore, studying the trade-offs and synergy between ecosystem services can better guide production and life practice and bring great benefits to human life.

With the rapid development of social economy, the change of natural environment, the ability of human beings to transform nature and the intensity of obtaining natural resources, environmental damaging behavior of humans occurs frequently, bringing great damage to the ecosystem, and the ecosystem service benefits are also reduced. In this context, research on ecosystems has attracted the attention of many international scientists, and the analysis and evaluation of ecosystems have gradually become a research hotspot in ecology [7–11]. In 1997, *Nature's Services: Societal Dependence on Natural Ecosystems* was published, regarded as a landmark work in the field of ecosystem services, making more scholars have a keen interest in the field of ecosystem services and begin to focus on research on ecosystem services [1]. Subsequently, Costanza [9], Millennium Ecosystem Assessment Plan (MA) [5] has been comprehensively analyzed by ecosystem services from different perspectives of different disciplines, functions, processes and internal connections. In recent years, people have paid more attention to the economic benefits brought by ecosystem services but ignored the most basic and important ecological benefits, which made the imbalance between the supply of services, regulatory services, support services and cultural services in the ecosystem, and the trade-off relationship showed the trend of gradual deterioration [12, 13].

In recent years, many Chinese scholars have also joined the ranks of studying ecosystem services and using various research methods to explore the relationship between Chinese ecosystem services found out their drivers, provided theoretical knowledge and scientific basis for government decision-making, and made great contributions. Current research methods for ecosystem service trade-offs mainly explore the relationship between services using rose graph analysis, production possibility boundary methods [14], correlation functions in R language [15], correlation analysis, linear and power function regression [16]. Some scholars, based on remote sensing images, explore the number of regional land-use changes, spatial and temporal differences; estimate ecosystem service index; and explore the trade-offs between ecosystem services [15]. These methods for regional ecological system service trade-off theory research laid a solid foundation and also help the national government to make major decisions.

Guangxi's southwest region mainly includes Nanning, Baise, Chongzuo, Qinzhou, Beihai and Fangchenggang. The region is Guangxi karst and karst landform distribution. The region as the southwest of China ecological security barrier area and core area for poverty alleviation, which is the national key development of Beibu Gulf economic zone ecological hinterland, is also the 21st-century Maritime Silk Road, the Silk Road Economic Belt and the organic important connection point of new western land and sea access. However, this region is a contiguous poverty-stricken area in southwest Guangxi, and a concentrated area of most poverty-stricken counties at the state level. Regions, rocky desertification and ecosystem structure stability are poor. In view of this, this study in southwest Guangxi as the research area and its characteristics develops the karst region ecosystem service trade-off evaluation model, evaluates the ecosystem service balance. The research results can provide technical support for the quantitative assessment of national ecological engineering governance measures, in order to promote the optimization of karst regional ecosystem services and improve regional sustainable development.

2. Study area overview

In this paper, southwest Guangxi was selected as the study area, southwest of Guangxi Zhuang Autonomous Region, the north bank of the Beibu Gulf, geographically located between 104°28′–109°56′E and 20°26′–25°07′N (**Figure 1**). The study area includes six urban areas: Baise, Nanning, Chongzuo, Qinzhou, Fangchenggang and Beihai, with the total area of 95,661.65 km^2, accounting for about 40% of the total area of Guangxi. In the study area, karst landform development is relatively typical as a geological formation. The landform assemblage form is mainly peak cluster depressions, peaks forest valleys, with steep terrain and large mountain slopes. The overall trend is tilted northwest to southeast. The overall mountain elevation of the study area is between −18 and 2036 m.

The study area belongs to the subtropical monsoon climate zone, characterized by short winter and long summer, high precipitation and long sunshine time. Its average annual temperature is between 19.81 and 23.09°C. The annual rainfall in the study area is abundant, obviously affected by the monsoon climate. Annual rainfall is between 87.71

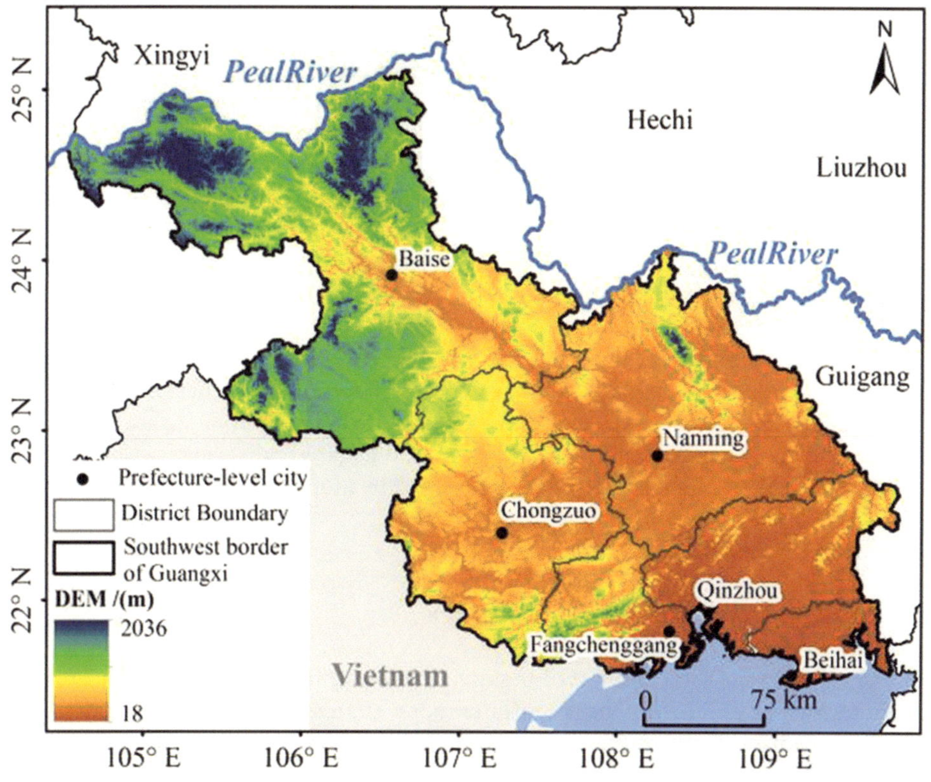

Figure 1.
Geographical location of southwest Guangxi.

and 214.01 mm, among which the rainfall is mainly concentrated in April–October. The area has strong sunshine, and the range of annual effective cumulative temperature is between 6803.32 and 8343 $J/m^2 \cdot a$, with relatively abundant heat, providing sufficient thermal energy conditions for plant growth. The landform of southwest Guangxi is complex. Under conditions of good subtropical climate, the region is rich in plants, with diverse vegetation types, such as evergreen broad-leaf forest, deciduous broad-leaf forest and shrub forest, and mangrove forests are also distributed in the southern coastal areas.

3. Materials and methods

3.1 Data collection and process

The main data used in this study are remote sensing data, meteorological data and land use type data.

1. Remote sensing data were obtained from NASA 2000 to 2018 (h27v06 and h28v06 MODIS13Q1) data products, a terrestrial level 3 standard data product with a temporal resolution of 16 days and a spatial resolution of 250 m × 250 m. Image correction, splicing, reprojection and clipping were preprocessed with MRT and ArcGIS software, and finally the maximum value synthesis method of the spatial analysis tool was used in ArcGIS to obtain the monthly Normalized Difference Vegetation Index (NDVI) data from January to December of each year.

2. Meteorological data were obtained from the China Meteorological Science Data Sharing Service Network (http://cdc.cma.gov.cn), containing day-to-day data on air temperature, rainfall, solar radiation, evapotranspiration. The time span is 2000–2018 from January to December, mainly including 25 meteorological stations in and around southwest Guangxi. ArcGIS interpolated daily temperature and sporadic daily meteorological data to obtain daily meteorological grid data. Rainfall data, by image rotation of the data in ENVI and ArcGIS, coordinate correction, change of data units, clipping, raster to point, interpolation analysis, format conversion and other treatments, obtained the monthly precipitation data in the study area.

3. Land use data and DEM data were derived from the Geospatial Data Cloud (http://www.gscloud.cn/), with a spatial resolution of 30 m × 30 m. Land use data were generated by manual visual interpretation using ENVI5.1. Land use data include six types of cultivaaq8ted land, woodland, grassland, waterbody, construction land and unused land. The final DEM data was obtained by mosaic raw data and clipping.

3.2 Research methods

3.2.1 Net primary productivity (NPP) estimation

NPP refers to the total amount of organic dry matter produced by green plants per unit time and per unit area, whose spatiotemporal changes are mainly dependent on the complex interactions between vegetation, soil and climate [17]. As an important component of the surface carbon cycle, NPP is a major factor in determining ecosystem carbon sinks [18]. NPP has now become an indispensable indicator in studies of the impact of global changes on ecosystems [19]. There are many models to estimate

NPP, which can be divided into ecosystem ecological process model, climate-related statistical model and optical energy utilization model. The model used in calculating NPP is the CASA model improved by Zhu Wenquan based on previous research, which is more consistent with the regional scale estimation of vegetation NPP [19].

3.2.2 Water conservation estimation

This study uses the water production module of InVEST model to study water conservation in the southwest region of Guangxi. The water production module is based on the principle of water balance. The precipitation of the unit raster minus the actual evaporation precipitation yields the water production of the unit raster [20, 21]. Combining the calculation of various parameters based on the raster is conducive to analyzing the influencing factors leading to spatial heterogeneity in the calculation results. The calculation formula is as follows:

$$Y_{xj} = \left(1 - \frac{AET_{xj}}{P_x}\right) \bullet P_x \tag{1}$$

The Y_{xj} in the formula is the water yield (mm) of unit grid x class j land use/coverage type in the study area, AET_{xj} is the actual annual evapotranspiration amount (mm) of unit grid x class j land use/coverage type in the study area, P_x is the annual precipitation (mm) in unit grid x in the study area.

3.2.3 Soil conservation service model

Soil conservation (*SC*) service was estimated using the RUSLE model and corrected based on experimental data from the study area to obtain a computational model meeting the study area [22], with the following formula:

$$SC(x) = SC(p) - SC(r) = R \times K \times LS \times (1 - C \times P) \tag{2}$$

In the formula, $SC(x)$ indicates the soil conservation ($t \cdot hm^{-2} \cdot a^{-1}$) in the study area, $SC(p)$ indicates the potential soil erosion amount ($t \cdot hm^{-2} \cdot a^{-1}$), $SC(r)$ indicates the actual soil erosion amount ($t \cdot hm^{-2} \cdot a^{-1}$), R indicates the precipitation erosion force factor $MJ \cdot mm/hm^2 \cdot h \cdot a$), K indicates soil erodibility factor ($t \cdot h/MJ \cdot mm$), LS indicates the slope length-slope factor, C indicates surface vegetation management and coverage factors, P indicates soil conservation measures factors, the dimensionless number between 0 and 1.

3.2.4 Production possibility frontier (PPF)

PPF represents a combination of maximum production of two products that can be achieved under the conditions of established economic resources and production technology. PPF is assuming that the number of two products acts as the positive half-axis of X and Y axes, respectively. In the case of certain resources, assuming that the full resources are used to produce A products, then the A products can produce 60, with B products 0, and so on. If 50 A products are produced, then 22 B products are produced, corresponding point (50, 22). If 40 A products are produced, then 33 B products are produced, corresponding point (40, 33). If 32 A products are produced, then 39 B products are produced, corresponding point (32, 39). If 20 A products are produced, then 45 B products are produced, corresponding point (20, 45). If the A products of 0 are produced, then 50 B products are produced, corresponding point (0, 50). Thus obtain the PPF between the products A and B (the coordinate data of the aforementioned points are designed according to the definition of the PPF) [23]. As shown in **Figure 2**, the product B in the figure shows a downward trend with the increase of product A, which shows the trade-off relationship between product A and product B.

From the economic point of view, the production possibility boundary has the following characteristics:

- Resource limitations play a main role in product configuration;
- Any point on the boundary of the production possibility represents the maximum output combination, points within the boundary indicate that the output of the combination does not reach the maximum value, and points outside the boundary indicate that the output cannot be achieved under the existing technology;
- The production possibility boundary is not fixed, which will shift inward or outward migration with the input of economy and technology.

This paper cites the production possibility boundary curve for the ecosystem with fixed natural resources to quantitatively investigate the trade-off and synergy relationship between ecosystem services in southwest Guangxi. Take the two ecosystem services of vegetation NPP and water conservation in southwest Guangxi as an example to make production possibility boundaries. Pareto curves between the two services were calculated and produced, first calculating the inter-annual mean of vegetation NPP and water conservation data using ArcGIS's metastatistic tool, dividing the annual vegetation NPP data from the annual water conservation data to obtain a layer where each cell raster was the ratio of the corresponding geographical vegetation NPP and water conservation. The resulting ratio layer, average annual vegetation NPP and annual water conservation layer were sampled and then exported as an Excel table.

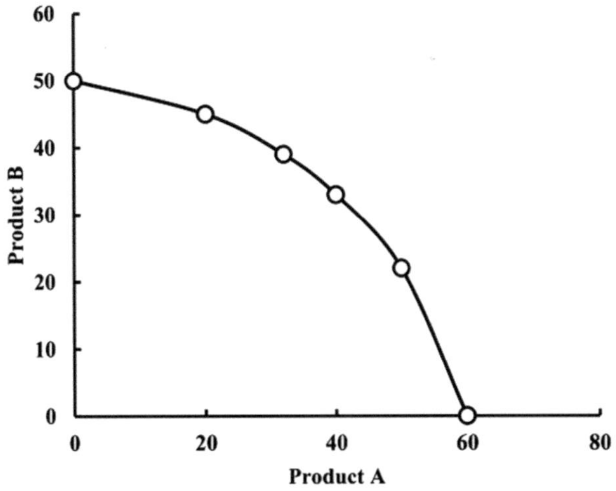

Figure 2.
Production possibility frontier (PPF) curve.

In the table, the vegetation NPP and water conservation ratio were sorted in ascending order, and the annual vegetation NPP and water conservation were summed in proper order, and then the Pareto efficiency curve was drawn according to the results.

4. Result analyses

4.1 Inter-annual changes of different ecosystem services in southwest Guangxi

4.1.1 Inter-annual change law of NPP in southwest Guangxi

The change trend of vegetation NPP in southwest Guangxi from 2000 to 2018 (**Figure 3**) showed the overall growth trend. The entire annual NPP fluctuated between 300 and 500 gC/m^2 a, with an annual growth rate of 3.781 gC/m^2 a in 2018, fluctuating upward trend in 2000–2006, gradual downward trend in 2006–2009 and 2009–2013 in 2019. The maximum appeared in 2018, showing a fluctuating upward trend in 2000–2006; a gradual downward trend in 2006–2009; a phenomenon of repeated alternation in 2009–2013.

In different lithological backgrounds, the overall trend of the mean in the study area was consistent with the non-karst region, generally consistent with the karst region, slightly from high NPP in 2005 and low in 2006 and 2016. The annual NPP in non-karst region was greater than the annual NPP in karst region. Among them, the NPP growth rate in non-karst regions was 5.852 gC/m^2 a, 3.188 gC/m^2 a in the karst region, indicating that the non-karst region in southwest Guangxi had a greater impact on NPP growth.

4.1.2 Inter-annual changes of water conservation in southwest Guangxi

Statistics on the changes in the mean value of water conservation area in southwest Guangxi in 2000–2018, as shown in **Figure 4**, shows the increase of water conservation in southwest Guangxi. The change trend of water conservation in karst landform and non-karst landform in 19 years is consistent with the overall study

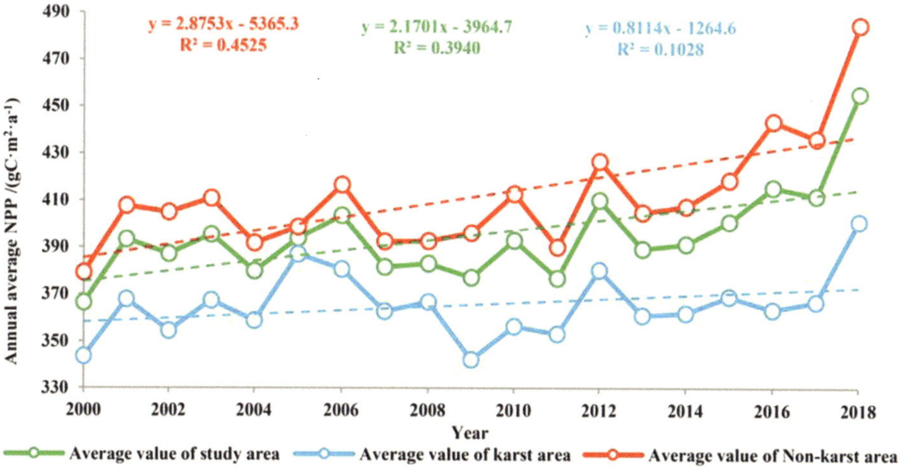

Figure 3.
The average change trend of NPP under different lithological backgrounds in southwest Guangxi from 2000 to 2018.

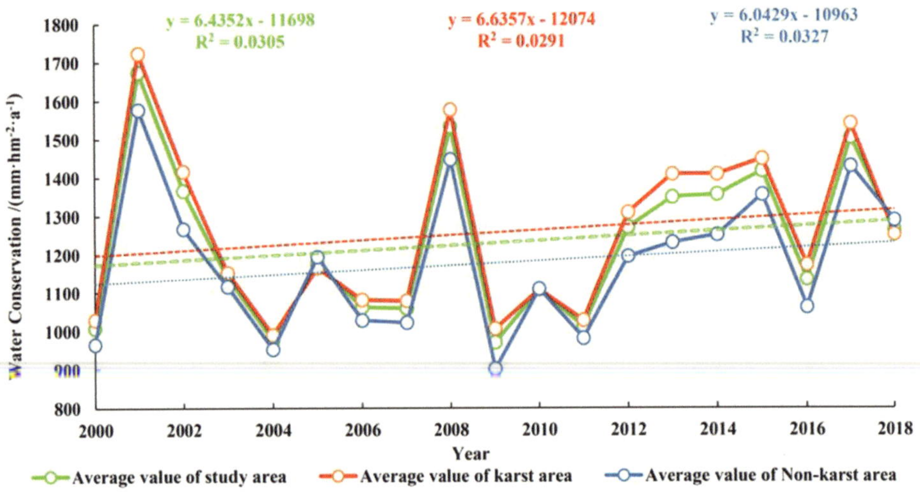

Figure 4.
Inter-annual mean variation of water conservation in southwest Guangxi.

area as a whole. Between 2000 and 2018, the water conservation in the study area was increased by 255.88 mm·hm^{-2}·a^{-1}, study area to 1230.16 mm·hm^{-2}·a^{-1}. Among them, water conservation in 2001 was the maximum in 19 years, higher than the mean in 444.76 mm·hm^{-2}·a^{-1}; water conservation in 2009 was the minimum in 19 years, lower than the mean in 260.18 mm·hm^{-2}·a^{-1}.

4.1.3 Inter-annual changes of soil conservation in southwest Guangxi

Through the ArcGIS statistical tool, the changes of karst landforms, non-karst landforms and soil retention in the whole region in different years were shown in **Figure 5**. The analysis found that the soil retention in southwest Guangxi was upward, with the soil retention in the karst landform more significant, and the change rate was that the soil retention (4.77/a) was greater than the whole region (2.23/a), and greater than the non-karst soil retention (0.93/a), in which the soil retention in the non-karst landform was generally consistent with that in the whole region. During 2000–2018, the range of soil retention variation to 65.38–96.88 t·hm^{-2}·a^{-1}, collectively increased 22.73 t·hm^{-2}·a^{-1}, with a mean of 79.19 t·hm^{-2}·a^{-1}. In 2016, soil retention fell to a minimum (65.38 t·hm^{-2}·a^{-1}), reached a maximum (96.88 t·hm^{-2}·a^{-1}) in 2017. Therefore, the soil retention maintained a rapid increase from 2016 to 2017. Overall, soil retention in the study area was wavy. 2000–2002, soil retention increased 22.46 t·hm^{-2}·a^{-1}; 2005–2007, soil retention reduced 20.11 t·hm^{-2}·a^{-1}, with its reduction rate 6.70/a; 2015–2018, soil retention showed a change in decrease to rise, and 2015–2016, soil retention reduced 17.26 t·hm^{-2}·a^{-1}, which was the worst decline in soil retention in 19 years, after increased change.

4.2 The spatial distribution pattern of ecosystem services in southwest Guangxi

4.2.1 The average annual NPP spatial distribution pattern in southwest Guangxi

According to the spatial distribution map of inter-annual average vegetation in southwest Guangxi from 2000 to 2018 (**Figure 6**), the average NPP of vegetation in southwest

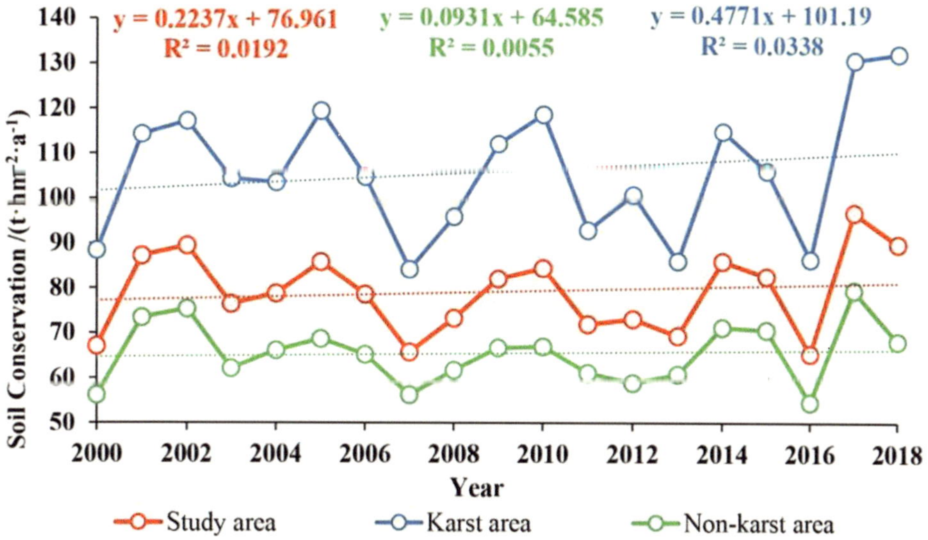

Figure 5.
Inter-annual changes of soil conservation in southwest Guangxi.

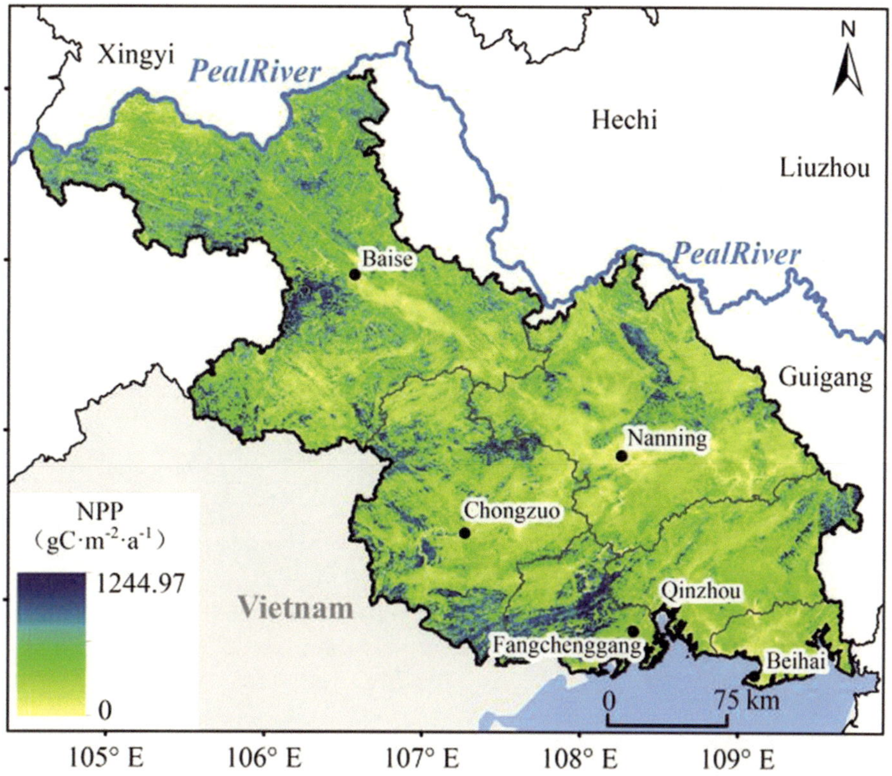

Figure 6.
Inter-annual distribution of vegetation NPP in southwest Guangxi from 2000 to 2018.

Guangxi was 394.95 gC/m^2 a in 19 years, decreasing from east to west and from northwest to southeast. The differences in the spatial distribution of the vegetation NPP in different areas in southwest Guangxi are also more obvious. The vegetation of Shiwan mountains in the middle of Fangchenggang city and Dawanglin mountains in the western marginal area of Baise City has high annual NPP. Due to the lush vegetation of Shiwan and Dawangling mountains, and other forest areas, less human activity and high vegetation productivity, the average annual NPP of these two prefecture-level cities is Fangchenggang city 490.18 gC/m^2·a and Baise city 437.73 gC/m^2·a, respectively. The larger range of NPP mean in southwest Guangxi is around 300–1000 gC/m^2·a, mainly concentrated in the northwest, west and south of southwest Guangxi, and is also distributed in parts of the east. Nanning, distributed in southwest central Guangxi, and Beihai, in southeastern Guangxi, are greatly affected by human activities, with urban expansion, low vegetation coverage and small productivity. The annual average of these two prefecture-level cities is Nanning city 330.13 gC/m^2·a and Beihai city 253.40 gC/m^2·a, respectively.

4.2.2 The average annual spatial distribution pattern of water conservation in southwest Guangxi

By calculating the inter-annual mean of water conservation for each raster in the study area, the raster plot (**Figure 7**) can analyze the spatial distribution characteristics of water conservation in southwest Guangxi. 2000–2018 inter-annual mean of water conservation was between 34.60 and 2099.46 mm·hm^{-2}·a^{-1}, with a mean of 1230.16 mm·hm^{-2}·a^{-1}.

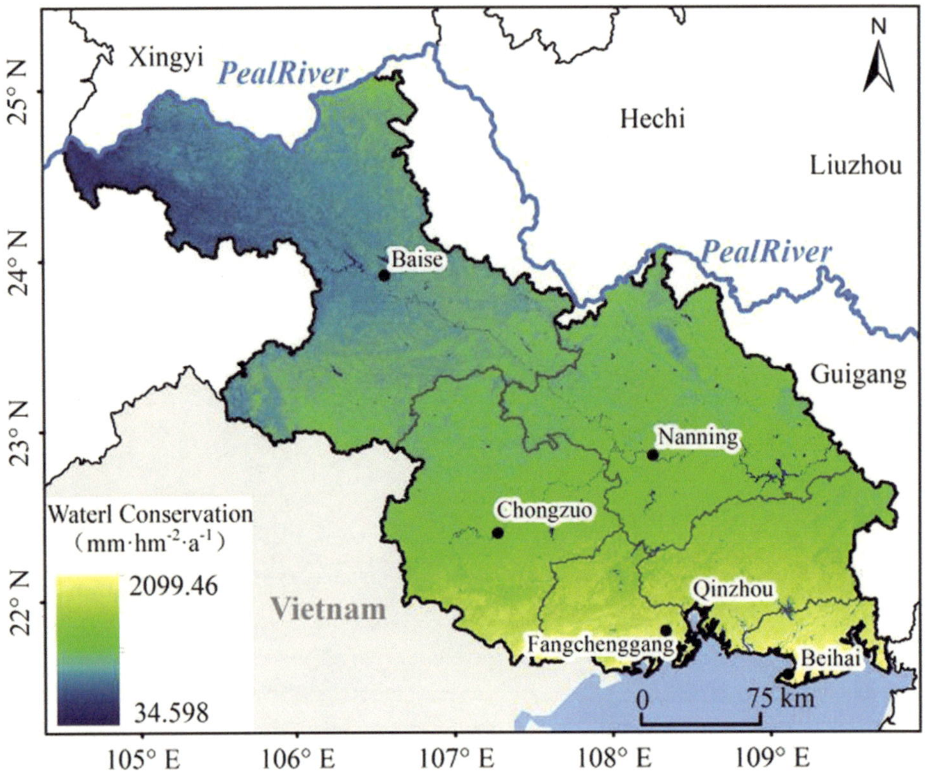

Figure 7.
Inter-annual mean distribution of water conservation mass in southwest Guangxi.

Overall, the high-value area of water conservation is in the south of the study area, and the low-value area is in the north of the study area. Its spatial changes show a decreasing trend from the south to the north. Analysis found that the distribution of water conservation at all levels in the research area is relatively regular, and the higher value area is concentrated in Ningming county, Fusui county of Chongzuo city and Yongning County, Heng county of Nanning city and other some areas, accounting for 21.93% of the total area, while the high value area accounted for 14.33% of the area, concentrated in Fangchenggang city, Qinzhou city and the southern area of Beihai city. Qinbeifang area is located near the sea, with more extensive fisheries. More artificially fishing farms established, so to some extent, it can be regarded as the regional water conservation capacity, and the precipitation in the research area is mostly concentrated in the southern region. The median value area was concentrated in the central region of the study area, mainly in Chongzuo City and Nanning City, accounting for the vast majority of the area, and also in the eastern cities of Baise city. Low and lower value areas collectively accounted for 30% of the total study area, with the median value (33.74%) being the most, followed by lower value (23.56%), higher value (21.93%) and the lowest.

4.2.3 The average annual spatial distribution pattern of soil conservation in southwest Guangxi

Soil conservation is influenced by factors such as topography, soil nature and precipitation erosion force, showing spatially different spatial regional differentiation characteristics. The inter-annual mean was calculated from soil retention from 2000 to 2018 by the ArcGIS software to obtain **Figure 8**. As can be seen from the figure, the range of soil retention was 0–2074.04 $t{\cdot}hm^{-2}{\cdot}a^{-1}$, with an average 79.18 $t{\cdot}hm^{-2}{\cdot}a^{-1}$, standard deviation of 104.77. In the inter-annual spatial distribution map of soil conservation, the high values were distributed in the west of southwest Guangxi and scattered, but the Shiwan mountains in Fangchenggang city are relatively concentrated, because the vegetation coverage of Shiwan mountains is high, which plays a large role in maintaining the soil. The middle of the study area was a low-value distribution, such as the central residential living zone of Baise City, the southwest of Nanning City and the vast area of Beihai City, which indicated a large connection between human activity and soil conservation.

4.3 Ecosystem service trade-off in the southwest Guangxi

4.3.1 The trade-off relationship between vegetation NPP and water conservation

From the production possibility boundary curve of vegetation NPP and water conservation (**Figure 9**), there is an obvious synergy relationship between vegetation NPP and water conservation in southwest Guangxi, and one of the ecosystem services will be improved, and the other ecosystem services will also be improved. Analyzing the relationship between vegetation NPP and water conservation, as in **Figure 9**, vegetation NPP increased by 2,003,924 $gC{\cdot}hm^{-2}{\cdot}a^{-1}$ from point a to point b, water conservation grew by 5,713,864 $mm{\cdot}hm^{-2}{\cdot}a^{-1}$, and vegetation NPP increased by 1,873,433 $gC{\cdot}hm^{-2}{\cdot}a^{-1}$ from b to point c, water conservation grew with 8,306,689 $mm{\cdot}hm^{-2}{\cdot}a^{-1}$. Comparing the two ecosystem services of a–b interval and b–c interval, the change in a–b stage is relatively slow, indicating that water conservation with the change of vegetation NPP began to increase slowly, and then increases gradually. From this change trend, the growth of vegetation NPP will cause an increase in water conservation, so it shows that vegetation NPP and water conservation in southwest Guangxi show a synergistic relationship.

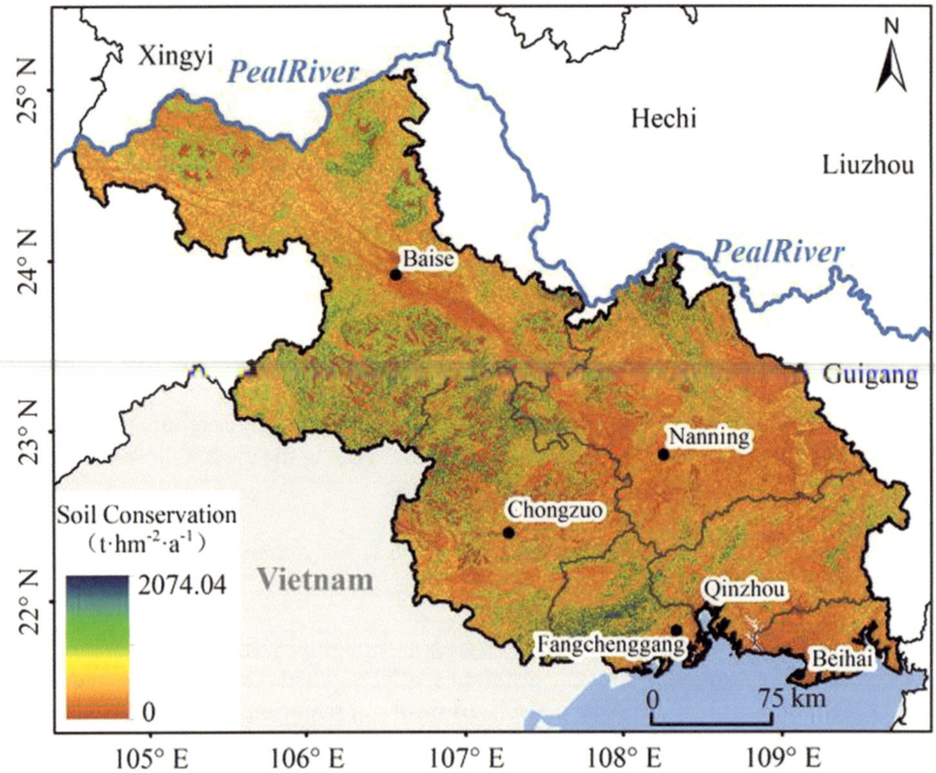

Figure 8.
Inter-annual mean distribution of soil conservation in southwest Guangxi.

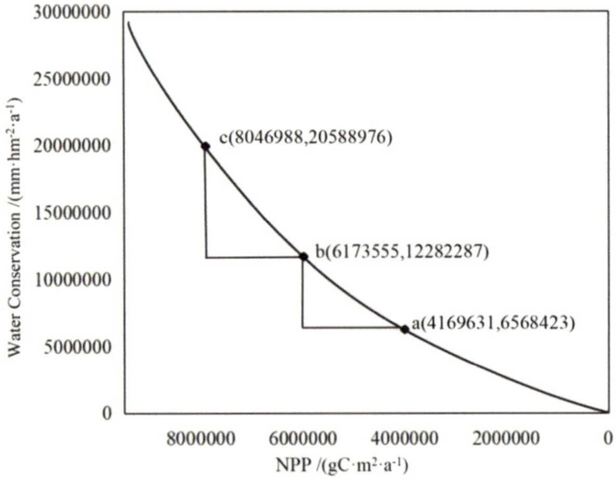

Figure 9.
The PPF curve of vegetation NPP and water conservation.

4.3.2 The trade-off relationship between vegetation NPP and soil conservation

Using ArcGIS and Excel software to count the values of vegetation NPP and make the production possibility boundary of the two services, obtain **Figure 10**,

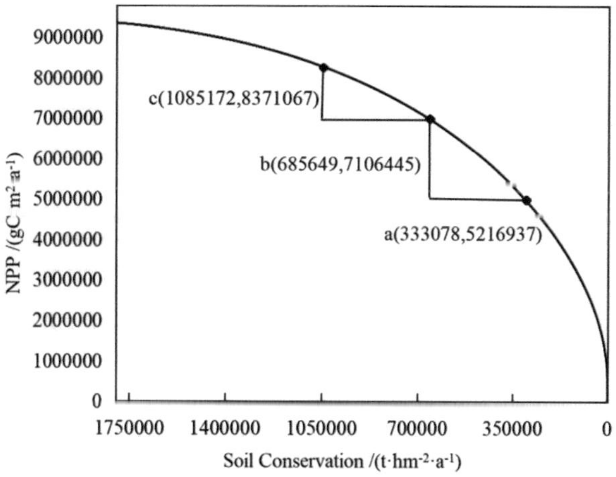

Figure 10.
The PPF curve of vegetation NPP and soil conservation.

which shows a synergistic relationship between the vegetation NPP in the southwest of Guangxi and soil retention. At the front end of the curve (i.e., the lower right corner of the curve), the soil retention remains at 0 and the vegetation NPP has gradually been increased. Three points were randomly selected on the production possibility boundary of vegetation NPP and soil retention to form two groups of change intervals. The change interval of analysis is as follows: by a–b interval, soil retention increment of 352,571 $t{\cdot}hm^{-2}{\cdot}a^{-1}$, vegetation NPP from 5,216,937 to 7,106,445 $gC{\cdot}hm^{-2}{\cdot}a^{-1}$, increment 1,889,508 $gC{\cdot}hm^{-2}{\cdot}a^{-1}$; by the interval from point b to c, vegetation NPP increased by 1,264,622 $gC{\cdot}hm^{-2}{\cdot}a^{-1}$, while the increment in soil retention 399,523 $t{\cdot}hm^{-2}{\cdot}a^{-1}$. Soil conservation shows the same trend with the change of vegetation NPP, indicating that one of the services is improved, and the other service also improved, showing a synergy relationship (e.g., more vegetation will mean more conserved soil because of less erosion and less exposed slopes).

4.3.3 The relationship between soil conservation and water conservation

By making the production possibility boundary between soil conservation and water conservation in southwest Guangxi (**Figure 11**), the trend of PPF curves with the one of vegetation NPP and soil conservation were generally similar, and when water conservation accumulated to 57,883 $mm{\cdot}hm^{-2}{\cdot}a^{-1}$, soil conservation only changed from 0 to 27. The production possibility boundary was still analyzed using three points into two change intervals, namely a–b and b–c intervals, and analyzing the change of service within the two intervals. In the a–b interval, the soil conservation increased from 144,106 to 533,141 $t{\cdot}hm^{-2}{\cdot}a^{-1}$, and water conservation from 12,714,849 to 21,002,026 $mm{\cdot}hm^{-2}{\cdot}a^{-1}$. In the b–c interval, increments of the soil conservation were 593,935 $t{\cdot}hm^{-2}{\cdot}a^{-1}$, and increments of water conservation were 5,625,893 $mm{\cdot}hm^{-2}{\cdot}a^{-1}$. Comparison of the a–b and b–c intervals revealed that the change rate of the a-b interval was somewhat faster than the b–c interval. On the whole, soil conservation and water conservation in southwest Guangxi will cooperate in 2000–2018. If soil conservation is improved, water conservation will also be improved.

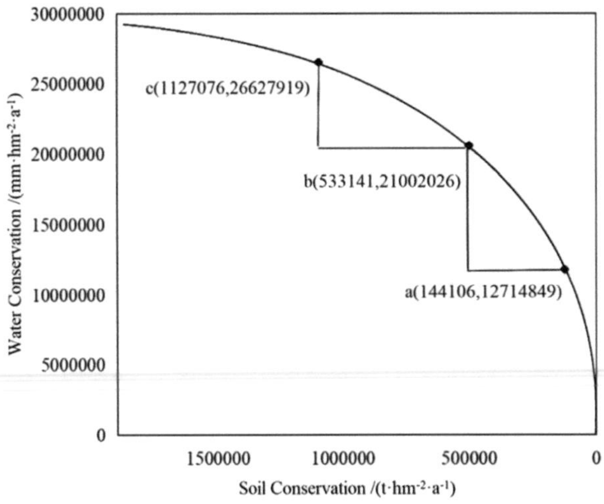

Figure 11.
The PPF curve of soil conservation and water conservation.

5. Conclusions

Vegetation NPP in southwest Guangxi increased from 2000 to 2018. The whole NPP annual average fluctuated between 300 and 500 gC/m^2 a, with an annual growth rate of 3.781 gC/m^2 a. The vegetation NPP average in southwest Guangxi was 394.95 gC/m^2 a in the past 19 years.

In 2000–2018, the vegetation NPP in southwest Guangxi generally showed a decreasing spatial distribution pattern from east to west and from northwest to southeast. In the past 19 years, the vegetation NPP has increased in most areas of southwest Guangxi. NPP shows a growing area of about 56,776 km^2, accounting for 59.35% of southwest Guangxi. The area of NPP, with a decreasing trend of about 38,887 km^2, accounts for 40.65% of southwest Guangxi.

The quality of water conservation and its value in the research area show a fluctuating trend, with the growth rate of water conservation quality 255.88 mm·hm^{-2}·a^{-1}, and the growth rate of water conservation value 97,725.88 yuan hm^{-2}·a^{-1}.

Soil conservation has increased in southwest Guangxi, with the obvious increase of soil retention in the karst landscape, and the change rate is ranked that the amount of soil retention (4.77/a) is greater than the whole region (2.23/a), and greater than the amount of non-karst landform (0.93/a), which is generally consistent with the soil retention in the whole region.

Vegetation NPP in southwest Guangxi and soil conservation, water conservation, soil conservation and water conservation are synergistic. In other words, to improve any of the three services, and other services will also be increased.

Acknowledgements

This work is supported by the National Natural Science Foundation of China (Grant No. 42061020), Natural Science Foundation of Guangxi Zhuang Autonomous Region (Grant No. 2018JJA150135), Guangxi Natural Science Foundation Youths

(Grant No. 2018JJB150094), Science and Technology Planning Project of Guangxi (Grant No. 2018 AD19040), Guangxi Key Research and Development Program (Grant Nos. AA18118038 and AB19110020), Science and Technology Department of Guangxi Zhuang Autonomous (Grant No. 2019 AC20088), High-level talent introduction project of Beibu Gulf University (Grant Nos. 2018KYQD19 and 2019KYQD28).

Conflict of interest

The authors declare no conflict of interest.

Author details

Yichao Tian[1,2,3] and Yongwei Yang[1*]

1 School of Resources and Environment, Beibu Gulf University, Qinzhou, China

2 Key Laboratory of Marine Geographic Information Resources Development and Utilization, Beibu Gulf University, Qinzhou, China

3 Guangxi Key Laboratory of Marine Disaster in the Beibu Gulf, Beibu Gulf University, Qinzhou, China

*Address all correspondence to: yangyongwei986@hotmail.com

References

[1] Daily GC. Nature's Services: Societal Dependence on Natural Ecosystems. Washington DC: Island Press; 1997

[2] Fu B, Zhou G, Bai Y, et al. The main terrestrial ecosystem services and ecological security in China. Advances in Earth Science. 2009;**24**(6):571-576

[3] Zheng H, OuYang Z, Zhao T, et al. The impact of human activities on ecosystem services. Journal of Natural Resources. 2003;**18**(1):118-126

[4] Qin K. Study on the Ecosystem Service Trade-off and Synergy between Guanzhong and Tianshui Economic Zone Based on Scenario Simulation. Xian: Shaanxi Normal University; 2016

[5] Board MA. Millennium Ecosystem Assessment. Washington, DC: New Island; 2005

[6] Gonzalez-Ollauri A, Mickovski SB. Providing ecosystem services in a challenging environment by dealing with bundles, trade-offs, and synergies. Ecosystem Services. 2017;**28**(12):261-263

[7] Board MEA. Millennium ecosystem assessment—Ecosystems and human well-being: Wetlands and water synthesis. Physics Teacher. 2005;**34**(9): 534-534

[8] Constanza R, Darge R, de Groot R, et al. The value of the world's ecosystem services and natural capital. World Environment. 1999;**25**(1):3-15

[9] Costanza R, Darge R, de Groot R, et al. The value of the world's ecosystem services and natural capital. Nature. 1997;**387**(6630):253-260

[10] Ruijs A, Wossink A, Kortelainen M, et al. Trade-off analysis of ecosystem services in Eastern Europe. Ecosystem Services. 2013;**4**:82-94

[11] Nagenda H, Reyers B, Lavorel S. Impacts of land change on biodiversity: Making the link to ecosystem services. Current Opinion in Environmental Sustainability. 2013;**5**(5):503-508

[12] Van Jaarsveld AS, Biggs R, Scholes RJ, et al. Measuring conditions and trends in ecosystem services at multiple scales: The Southern African Millennium Ecosystem Assessment (SAfMA) experience. Philosophical Transactions of the Royal Society, B: Biological Sciences. 2005;**360**(1454):425-441

[13] Barbier EB, Koch EW, Silliman BR, et al. Coastal ecosystem-based management with nonlinear ecological functions and values. Science. 2008;**319**(5861):321-323

[14] Yang X, Zhou Z, Li J, et al. Trade-offs between carbon sequestration, soil retention and water yield in the Guanzhong-Tianshui Economic Region of China. Journal of Geographical Sciences. 2016;**26**(10):1449-1462

[15] He M. Space and Temporal Difference Analysis of the Ecological Effect of Land Change in Yan'an City. Xian: Shaanxi Normal University; 2018

[16] Sun Y, Ren Z, Zhao S, et al. Spatial and temporal changing analysis of synergy and trade-off between ecosystem services in valley basins of Shaanxi Province. Acta Geographica Sinica. 2017;**72**(3):521-532

[17] Ewuketu L, Aaron-Hogan J, Cao M, et al. Precipitation influences on the net primary productivity of a tropical seasonal rainforest in Southwest China: A 9-year case study. Forest Ecology and Management. 2020;**467**(2):118153

[18] Nianpeng H, Yu Q, Wu L, et al. Carbon and nitrogen store and storage potential as affected by land-use in a *Leymus chinensis* grassland of northern China. Soil Biology and Biochemistry. 2008;**40**(12):2952-2959

[19] Zhao MS, Heinsch FA, Nemani RR, et al. Improvements of the MODIS terrestrial gross and net primary production global data set. Remote Sensing of Environment. 2004;**95**(2): 164-176

[20] Liu S. Wetland Ecosystem Services Evaluation Based on InVEST Model. Harbin: Harbin Normal University; 2019

[21] Liu J, Fu B, Zhang C, et al. Assessment of ecosystem water retention and its value in the upper reaches of Minjiang River based on InVEST model. Resources and Environment in the Yangtze Basin. 2019;**28**(3):87-95

[22] Wu G, Lin H, Zeng H. Responses of soil conservation function to ecosystem changes: An assessment based on RS and GIS in Changting County, Fujian Province. Acta Ecologica Sinica. 2017;**37**(1):321-330

[23] Stosch KC, Quilliam RS, Bunnefeld N, et al. Quantifying stakeholder understanding of an ecosystem service trade-off. Science of the Total Environment. 2018;**651**(2): 2524-2534

Chapter 7

Wooden Extra Stories in Concrete Block of Flats in Finland as an Ecologically Sensitive Engineering Solution

Markku Karjalainen, Hüseyin Emre Ilgın and Dennis Somelar

Abstract

This chapter examined the various stages and benefits of wooden extra stories from the perspective of Finnish housing and real estate companies through interviews with professionals involved in these projects. Key findings highlighted are as follows: (1) in the feasibility study, project planning primarily focuses on property condition and potential improvement targets as well as other considerations, for example, compliance with current regulations and parking arrangements; (2) in the project planning, application of extra stories is thoroughly examined, and construction costs, profits, and the sale of building rights are discussed; (3) in implementation planning, issues related to building rights, city plan change, and conditions of the company that manages the property play an important role; and (4) during construction, frequent information updates are made to residents regarding the site arrangements and the construction program. Wooden extra floor construction, which requires commitment, investment, and cooperation among the interested parties, has great potential in construction technology, contracting mechanisms, and ecological engineering solutions. It is believed that this chapter will increase the dissemination of wooden extra stories, thus contributing to the greater use of more sustainable materials in renovation projects and the ecologically sensitive engineering approaches to meet the challenges arising from climate change.

Keywords: timber/wood, wooden extra story, ecological engineering solutions, sustainable material, sustainability, building construction

1. Introduction

Climate change has reached a critical level [1, 2]. The probability of attributing global climate change to human factors is 95% and the risk of anthropogenic climate change requires the management of our operations [3]. In this sense, around 40% of the EU's total energy consumption and more than 35% of energy-related greenhouse gas emissions come from buildings [4]. The role of buildings, especially houses, in our lives during the COVID-19 crisis is more critical and the home has become the focal point of the daily life of millions of people in the EU, as more people started working

IntechOpen

from home. For example, some evidence-based surveys showed a large increase in the share of workers working from home compared to pre-crisis figures; from 30% in Canada to 70% in South Africa [5].

Especially at this point, renovation can turn crisis into opportunities to make our residential buildings suitable for a greener and digital society and to sustain economic recovery. To achieve 55% emission reductions in the 2030 climate target plan, the EU must reduce the greenhouse effect of buildings by 60% gas emissions, final energy consumption by 14%, and energy consumption for heating and cooling by 18% [6]. Moreover, emissions during building construction accounted for a total of 7% of Finland's greenhouse gas emissions in 2018, and emissions from the use of buildings account for 23% of the total [7]. Therefore, a focus on renovating the existing building stock to make it more energy-efficient and less carbon-intensive becomes urgent as an issue of ecological engineering in the context of circular economy and sustainable development [8, 9].

As in many European countries, Finland's building stock was mostly constructed before the 1990s, and this energy-poor stock often needs refurbishment to meet new building standards [10, 11]. This old stock, especially residential buildings built in the 1960s and 1980s (**Figure 1**), requires major renovations to approach a sustainable and carbon-neutral built environment [12, 13]. The renewal of building codes in Finland in recent years has contributed to accelerating the above transformation by aiming to make new construction methods increasingly more ecologically sound, sustainable, and at the same time more energy-efficient [14, 15].

Housing and real estate companies play an important role in building renovation that makes older buildings more sustainable, ecological, and energy-efficient [16]. The issue becomes even more important when it is considered that there are over 60,000 flats in Finland, where nearly half of the Finnish population lives [11]. Renovating an apartment by increasing accessibility and energy efficiency is a slow and costly process, requires a lot of capital and government subsidies, and the intensity of fieldwork distracts building occupants and residents living near the construction site [17–20].

In this context, extra story construction (**Figure 2**), which is advantageous in many aspects such as financial, environmental, and energy efficiency, stands out as a viable solution. It can cover the costs of property development and renovation to finance the necessary measures, directly increase the energy efficiency and indirectly support the energy renovation using the revenues from it, and finance the

Figure 1.
A residential complex in Finland built in the 1960s (source: Wikipedia/Tiia Monto).

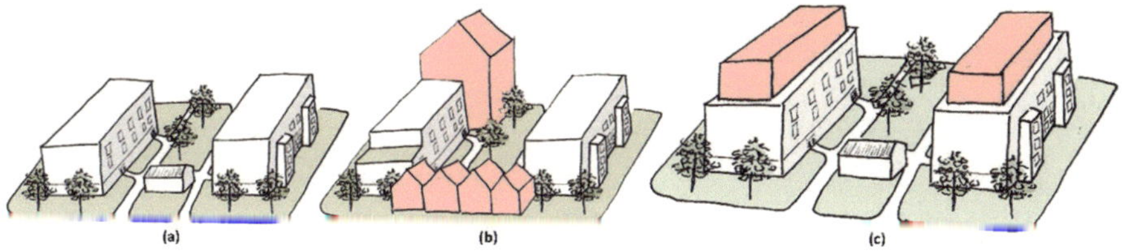

Figure 2.
Additional (story) construction: (a) basic; (b) additional construction; and (c) extra story construction.

construction of a retrofit elevator or new/extended balconies, thereby increasing the sustainability and living comfort [21–23].

Reinforced concrete-framed apartments built in Finland between the 1960s and 1980s allow the construction of extra stories, often using lightweight structures. Since these structures have flat roofs, they make it easier to add floors architecturally and technically. In addition, Finland's current fire regulations enable lightweight extra stories from materials such as wood [24].

The materials used in renovation should be renewable, recyclable, and long-lasting and their production should be sustainable and ecologically sound in nature, consuming minimum energy and producing as few emissions as possible as an ecological engineering solution [25, 26]. In this sense, timber is one of our best partners in tackling the climate crisis due to its potential eco-friendly properties, for example low-carbon emissions, and is at the forefront of addressing European climate policy [27–33]. Moreover, wood offers light prefabricated alternatives to meet the special design needs with its wide thermal insulation and size options [34, 35].

The reform of the legislation on building and construction also prepares for a transition toward low-carbon building. In the future, carbon reduction must be taken into account in the whole life cycle of a building, that is in new buildings and renovation and demolition [36]. In this case, extra story construction stands out as a sustainable and practical alternative.

To date, no study in the literature provides a comprehensive understanding of this sustainable approach, especially in housing facilities. In this chapter, the various phases, advantages, and disadvantages of the wooden extra story are examined from the perspective of housing and real estate companies by interviews. These interviews underlined the main points in the four stages of extra story construction—(1) feasibility study; (2) project planning; (3) implementation planning; and (4) construction.

It is worth noting here that as a recently matured discipline, ecological engineering or ecotechnology is a combination of applied ecology, environmental engineering, biotechnology, systems control, and complexity sciences and has a wide range of applications such as conservation and restoration of natural habitats [37, 38]. Ecological engineering has become an important tool, especially today, with the use of sustainable materials to tackle climate change challenges. Considering that more than one-third of energy-related greenhouse gas emissions originate from buildings, renovation of buildings with wooden extra stories will make a significant contribution to combating the climate crisis in the context of ecological engineering.

In this chapter, wood and timber refer to engineered wood products such as cross-laminated timber [(CLT) a prefabricated multi-layer EWP, manufactured from at least three layers of boards by gluing their surfaces together with an adhesive under pressure], laminated veneer lumber [(LVL) made by bonding together thin vertical softwood veneers with their grain parallel to the longitudinal axis of the section, under heat and pressure)], and glue-laminated timber (glulam) [(GL) made by gluing together several graded timber laminations with their grain parallel to the longitudinal axis of the section)].

2. Literature survey: wooden extra story construction

The terms extra floor, roof, or elevation are used when the roof shape of the building changes, the height increases, and the number of floors of the buildings increases. As in many examples in Finland (**Figure 3**), one of the effective ways to improve the property is to change the use of buildings (**Figure 4**) as well as to construct extra floors [39]. Moreover, extra stories are currently being built by using modern construction methods with lightweight prefabricated elements, and buildings constructed in Finland in the 1960s and 1980s can usually support one or two extra stories [40].

Extra story construction has many benefits such as [41]—(i) from an environmental point of view, as an ecologically sensitive engineering solution, renovation and improvement operations with extra stories were more than 20% lower in carbon footprint compared to demolition and new construction [39], thus it increases the income of the property owners, resulting in a beneficial development of the building stock; (ii) from a financial standpoint, its revenues can be used to finance the renovation of existing property, such as the renovation of an elevator (**Figure 5**), to improve the

Figure 3.
Extra story construction examples from Finland.

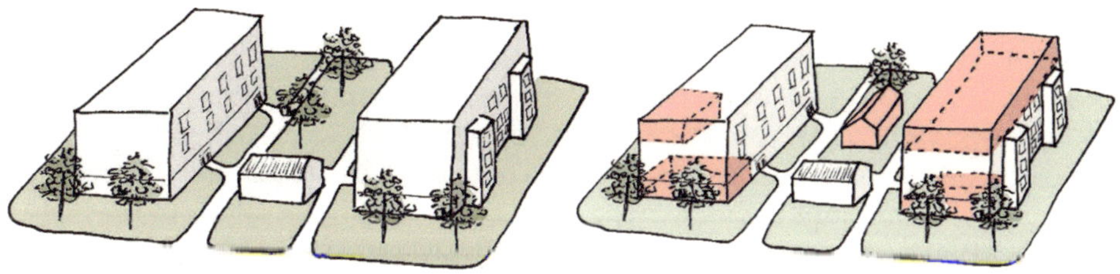

Figure 4.
Roof and ground floor renovations and building usage changes to increase efficiency. The building's internal storage facilities have been moved to a new outbuilding.

Figure 5.
A retrofit elevator can affect the architectural appearance of the building as well as accessibility.

building's accessibility and commercial conditions; (iii) in terms of energy efficiency, extra story construction does not significantly increase the overall energy consumption of the building, although it significantly increase the total floor area and as passive energy efficient structures, it can significantly improve the energy efficiency of old buildings, especially if the upper floors have not been renovated for a long time; (iv) from esthetical point of view, it can significantly affect the architectural features of the building and improve the facade appearance; and (v) from a social (sustainability) point of view, that is understanding people's needs and desires (e.g., [42, 43]), extra story construction that meets the demands of residents, where Finnish suburban residents generally have a positive attitude toward this sustainable solution [44].

On the other hand, according to some studies in the literature (e.g., [45–47]), the lack of cost-competitiveness of wood compared to conventional materials such as steel and concrete can be considered a disadvantage of wood in the construction of extra stories.

The materials used in renovations should be renewable, recyclable, and long-lasting and their production should consume only a minimum amount of energy and produce as few emissions as possible to ensure the sustainable ecology of the business [48]. Studies indicate that wood-based products are associated with far fewer greenhouse gas emissions over their lifetime than traditional building materials such as concrete and steel [49–52]. While concrete production accounts for around 8% of world CO_2 emissions [53], wood construction represents a lower concrete energy consumption compared to steel and concrete production [29]. Buildings using steel and concrete contain and consume 12- and 20% more energy, respectively than wooden buildings [27]. Moreover, products made from sustainably sourced wood replace other fossil-intensive substitutes, like concrete and steel [54–59].

Prefabricated timber frame offers lightweight and customizable solutions with multiple thermal insulation options to meet specific design needs [35, 60–62]. Prefabrication is highly efficient [63], high quality [64], and low cost [65], contributing to more than 50% minimization of construction waste and 70% wood formwork savings [66–70], compared to traditional construction methods. While prefabrication offers environmental friendly solutions with low carbon-emission and high utilization technology, it also reduces carbon emissions during transportation and ultimately contributes to an ecologically sound engineering approach [71, 72].

3. Research methods

This chapter was conducted in the form of a literature review including international peer-reviewed journals and similar research projects, and interviews professionals involved in the construction of extra stories during the PKRKP project (construction of wooden apartment buildings for growth in Pirkanmaa between 2019 and 2021). This project includes the city of Metsäkeskus and Tampere, as well as 14 municipalities and 15 companies in the Pirkanmaa region.

It is worth mentioning here that this chapter focused on housing and real estate companies. As responsible parties, they play a critical role in renovating and maintaining old apartments in Finland [16], where around half of the Finnish population lives [73]. In this study, semi-structured interviews were moderated as the process allowed for interviewer and interviewee interactions and various views encouraged the generations of new ideas beyond those originally explored [74].

Interviews were conducted with seven construction professionals who are part of a wooden extra story construction project and a housing company. The participants were—1. property development manager (from contractor side); 2. extra story construction consultant (from developer side) 3. city planner (from municipality); 4. project director/city planning (from municipality); 5. project director/city planning (from municipality); 6. CEO, associate, architect (from architectural design office); 7. CEO, associate, housing manager (from housing company side); and 8. member of the board of directors of a housing company (from housing company).

During interviews, the board of directors/representatives of the housing company were asked about mapping the progress of the project and the current situation, project burdens, the outcome of the project. The planning authority (city planner) was asked about city strategy and city plan change. The architect/chief designer was asked about project progress—contacting the housing company and combining extra story construction and building renovation. The housing manager was asked about project progress, and combining extra story construction, and building renovation. Extra story construction consultant was asked about project progress—contacting the housing company. Builders/contractors were asked about their organizational background. All participants were asked to report whether they had any other thoughts other than the specified themes in the other remarks section.

Interviews underlined the main points in the four stages of extra story construction—(1) feasibility study; (2) project planning; (3) implementation planning; and (4) construction, as summarized below.

4. Interviews on wooden extra story construction

The interviews highlighted the following key findings regarding four main phases of extra floor construction.

4.1 Feasibility study

Features for the feasibility study phase, in which the construction conditions of the extra stories are scrutinized and professionals in the construction and real estate sectors are contacted, are as follows:

City planning	It is important to determine the possibilities of permits or deviations from the city plan. These issues include how the extra floors relate to the surrounding buildings and the shading effect.
Parking space	The amount of parking required for the area is determined on a case-by-case basis by city authorities.
Information flow	Information provided in the drawings may vary from site to site. Completion of building drawings can be costly if the information is sought from the city's building inspectorate.
Load-bearing capacity	The housing or real estate company must compile existing drawings of the building to inspect the structure. Load-bearing capacity can be calculated directly from the drawings but also may require structural analysis.
Existing regulations	New regulations may affect operations; for example, if extra stories are added to an apartment without an elevator, retrofit elevators are needed.

4.2 Project planning

Themes for the project planning phase, in which the conditions of the project are determined, considerations are as follows:

Sale of building right	When an additional extension building right is sold to a third party, a recourse fee is allocated to the shares; also, a separate compensation may be determined which may include the costs and extra stories used in the project.
Building codes	They affect the implementation of the project, regardless of the building material from which the extra stories are constructed. For example, Finnish fire codes must be considered at an early stage, especially when designing a wooden extension where the codes allow the construction of two wooden extra stories.

4.3 Implementation planning

In the implementation planning phase, in which measures are taken to increase the building right, considerations are as follows:

City plan change	The permission of the landowner and material showing the effect of the extra story on the immediate surroundings of the property is required. Such studies may include elevation plans and drawings of their suitability in the immediate environment, a site drawing, and a building stock inventory.
Building right	In case the building right of the extra stories is sold to an external developer, a tender can be made after the building right has been added to the land.
Terms and conditions	The company managing the property to be extended vertically may set conditions for the sale of the building right. The terms and conditions can protect the interests of the company.

4.4 Construction

In the construction phase, issues are as follows:

Effective information flow	Residents and stakeholders should be kept informed of construction site progress, schedules, and potential times when construction significantly impairs residents' comfort of life.
Appointing a representative	In planning critical milestones, such as demolition work, it is an effective method for a housing or real estate company representative to attend site meetings and negotiate available times for work to minimize inconvenience from construction to residents.

5. Conclusion

Due to the lack of research on extra floor construction, it was not possible to provide a discussion of the similarities and dissimilarities of the Finnish practice with the applications in other regions. This chapter analyzed the different stages and advantages of wooden extra story construction from the standpoints of Finnish housing and real estate companies through interviews with professionals who have worked in these projects.

The key points from this chapter on the four phases of the wood extra story can be summarized as follows—(a) in the feasibility study, project planning primarily focuses on property condition and potential improvement targets, as well as other considerations, for example compliance with current regulations and parking arrangements; (b) in project planning, application of extra stories, is thoroughly examined, construction costs, profits and the sale of building rights are discussed; (c) in implementation planning, issues related to building rights, city plan change, and conditions of the company that manages the property play an important role; and (d) during construction, frequent information updates are made to residents regarding the site arrangements and the construction program.

Wooden extra story construction, which necessitates commitment, investment, and cooperation between interested parties, has great potential in construction technology, commissioning mechanisms, and ecological engineering solutions. Additionally, this sustainable approach with prefabricated timber solutions has many benefits in terms of economic, energy efficiency, esthetics, and environmental. In this sense, it is thought that this study will increase the popularity and prevalence of wooden extra stories, as in the case of Finland, thus contributing to the greater use of more sustainable materials in renovation projects and contributing to the ecologically sensitive engineering approaches to meet the challenges arising from climate change.

Author details

Markku Karjalainen, Hüseyin Emre Ilgın* and Dennis Somelar
Tampere University, Tampere, Finland

*Address all correspondence to: emre.ilgin@tuni.fi

References

[1] Liu Z, Chen Y. Impacts, risks, and governance of climate engineering. Advances in Climate Change Research. 2015;**6**(3-4):197-201. DOI: 10.1016/j.accre.2015.10.004

[2] Prashant K. Climate change and cities: Challenges ahead. Frontiers in Sustainable Cities. 2021;**3**(5). DOI: 10.3389/frsc.2021.645613 ISSN=2624-9634

[3] IPCC. Climate Change 2013: The Physical Science Basis. Contribution of Working Group I to the Fifth Assessment Report of the Intergovernmental Panel on Climate Change. Cambridge and New York: Cambridge University Press; 2013. Available from: https://www.sciencedirect.com/science/article/pii/S1876610217348592

[4] European Commission. New Rules for Greener and Smarter Buildings Will Increase Quality of Life for All Europeans [Internet]. 2019. Available from: https://ec.europa.eu/info/news/new-rules-greener-and-smarter-buildings-will-increase-quality-life-alleuropeans-2019-apr-15_en [Accessed: 01 October 2021]

[5] ILO-OECD. The Impact of the COVID-19 Pandemic on Jobs and Incomes in G20 Economies [Internet]. 2020. Available from: https://www.ilo.org/wcmsp5/groups/public/---dgreports/---cabinet/documents/publication/wcms_756331.pdf [Accessed: 01 October 2021]

[6] Communication from The Commission to The European Parliament, The Council, The European Economic and Social Committee and the Committee of The Regions, EUROPEAN COMMISSION, Stepping up Europe's 2030 Climate Ambition Investing in a Climate-neutral Future for the Benefit of Our People; Brussels, 17.09.2020, COM (2020) 562 final. 2020

[7] Raivio T, Laine A, Klimscheffskij M, Heino A, Lehtomäki J. Vähähiilinen rakennusteollisuus 2035, Osa 4—Rakennusteollisuuden ja rakennetun ympäristön vähähiilisyyden tiekartta 2020-2035-2050. Rakennusteollisuus RT ry. s.9 [Internet]. 2020. Available from: https://www.rakennusteollisuus.fi/globalassets/ymparisto-ja-energia/vahahii-lisyys_uudet/rt_4.-raportti_vahahiilisyyden-tiekartta_lopullinen-versio_clean.pdf [Accessed: 01 October 2021]

[8] Mercader-Moyano P, Esquivias PM. Decarbonization and circular economy in the sustainable development and renovation of buildings and neighbourhoods. Sustainability. 2020;**12**(19):7914. DOI: 10.3390/su12197914

[9] Schönborn A, Junge R. Redefining ecological engineering in the context of circular economy and sustainable development. Circular Economy and Sustainability. 2021;**1**:375-394. DOI: 10.1007/s43615-021-00023-2

[10] Simson R, Fadejev J, Kurnitski J, Kesti J, Lautso P. Assessment of retrofit measures for industrial halls: Energy efficiency and renovation budget estimation. Energy Procedia. 2016;**96**:124-133

[11] European Commission. Long-Term Renovation Strategy 2020-2050 Finland, Report According to Article 2a of Directive (2010/31/EU) on the Energy Performance of Buildings, as Amended by Directive 2018/844/EU [Internet]. 2020. Available from: https://ec.europa.eu/energy/sites/default/files/documents/

fi_2020_ltrs_en.pdf [Accessed: 01 September 2021]

[12] Kaasalainen T, Huuhka S. Homogenous homes of Finland: Standard flats in non-standardized blocks. Building Research and Information. 2015;**44**(3)

[13] Hirvonena J, Jokisaloa J, Heljo J, Kosonena R. Towards the EU emissions targets of 2050: Optimal energy renovation measures of Finnish apartment buildings. International Journal of Sustainable Energy. 2018;**38**(7):1-24

[14] Energy Policies of The International Energy Agency (IEA) Countries. Finland 2018 Review [Internet]. 2018. Available from: https://www.connaissancedesenergies.org/sites/default/files/pdf-actualites/situation_energetique_de_la_finlande.pdf [Accessed: 01 October 2021]

[15] Kuittinen M, Häkkinen T. Reduced carbon footprints of buildings: New Finnish standards and assessments. Buildings and Cities. 2020;**1**(1):182-197

[16] KTI Finland. The Finnish Property Market [Internet]. 2019. Available from: https://kti.fi/wp-content/uploads/The-Finnish-Property-Market-2019.pdf [Accessed: 01 October 2021]

[17] Official Statistics of Finland (OSF). Homeowners and Housing Companies Repaired by EUR 6.0 Billion in 2019 [Internet]. 2019. Available from: http://www.stat.fi/til/kora/2019/01/kora_2019_01_2020-06-11_tie_001_fi.html%20 [Accessed: 01 October 2021]

[18] Mills B, Schleich J. Residential energy-efficient technology adoption, energy conservation, knowledge, and attitudes: An analysis of European countries. Energy Policy. 2012;**49**:616-628

[19] Janda KB. Building communities and social potential: Between and beyond organizations and individuals in commercial properties. Energy Policy. 2014;**67**:48-55

[20] D'Oca S, Ferrante A, Ferrer C, Pernetti R, Gralka A, Sebastian R, et al. Technical, financial, and social barriers and challenges in deep building renovation: Integration of lessons learned from the H2020 cluster projects. Buildings. 2018;**8**:174

[21] Soikkeli A. Additional floors in old apartment blocks. Energy Procedia. 2016;**96**:815-823

[22] Jurelionis A, Šeduikyte L. Multifamily building refurbishment process in Lithuania and other European countries. In: Proceedings of 10th International Conference Modern Building Materials, Structures and Techniques; 19-21 May 2010; Vilnius, Lithuania: Vilnius Gediminas Technical University Press; 2010. pp. 106-111

[23] Matic D, Calzada JR, Eric M, Babin M. Economically feasible energy refurbishment of prefabricated building in Belgrade. Energy Buildings. 2015;**98**:74-81

[24] Botici AA, Ungureanu V, Ciutina A, Botici A, Dubina D. Sustainable retrofitting of large panel prefabricated concrete residential buildings, in CESB 2013 Sustain. In: Build. Refurb. Next Gener.; 26-28 June 2013; Prague; Czech Republic. Available from: https://www.scopus.com/record/display.uri?eid=2-s2.0-84925258549&origin=inward&txGid=36a90933056f1239c4fcdd8f211327c1&featureToggles=FEATURE_VIEW_PDF:1

[25] Myer F, Fuller R, Crawford RH. The potential to reduce the embodied energy in construction through the use of renewable materials, in ASA 2012: Building on knowledge, theory and practice: Proceedings of the 46th Annual

Conference of the Architectural Science Association, Architectural Science Association, Gold Coast, Qld., 2012. pp. 1-8

[26] Recycling Magazine. Construction Industry Progress Towards Sustainability with Renewable Materials [Internet]. 2020. Available from: https://www.recycling-magazine.com/2020/04/14/construction-industry-progress-towards-sustainability-with-renewable-materials/ [Accessed: 01 October 2021]

[27] CWC. Energy and the environment in residential construction. Sustainable Building Series No.1. 2007. Available from: https://cwc.ca/wp-content/uploads/publications-Energy-and-the-Environment.pdf

[28] Green M. The Case for Tall Buildings: How Mass Timber Offers a Safe, Economical, and Environmentally Friendly Alternative for Tall Building Structures. Canada: MGB Architecture and Design; 2012

[29] Wang L, Toppinen A, Juslin H. Use of wood in green building: A study of expert perspectives from the UK. Journal of Cleaner Production. 2014;**65**:350-361

[30] Bosman R, Rotmans J. Transition governance towards a bioeconomy: A comparison of Finland and the Netherlands. Sustainability. 2016;**8**(10)

[31] Karjalainen M. In: Lilja K, editor. Status and Possibilities of Timber Construction in Finland, 15 May 2017, Wood-Based Bioeconomy Solving Global Challenges. Helsinki: Ministry of Economic Affairs and Employment of Finland, Vol. MEAE 2/2017; 2017. pp. 35-39

[32] Karjalainen M. Multistory timber apartment buildings are becoming common in forested Finland—Results of resident surveys point the way to the future of multistory timber apartment buildings. In: Proceedings of 5th Annual International Conference on Architecture and Civil Engineering (ACE 2017); 8-9 May 2017. Singapore: Global Science & Technology Forum (GSTF); 2017. pp. 415-417

[33] Wood Building Programme. Finnish Ministry of Environment [Internet]. 2016. Available from: https://www.ym.fi/enUS/Land_use_and_building/Programmes_and_strategies/Wood_Building_Program/Wood_Building_Programme(47800 [Accessed: 01 October 2021]

[34] Gustavsson L, Joelsson A, Sathre R. Life cycle primary energy use and carbon emission of an eight-storey wood-framed apartment building. Energy Buildings. 2010;**42**:230-242

[35] Sandberg K, Orskaug T, Andersson A. Prefabricated wood elements for sustainable renovation of residential building façades. Energy Procedia. 2016;**96**:756-767

[36] Huuhka S, Vainio T, Moisio M, Lampinen E, Knuuttinen M, Bashmakov S, Köliö A, Lahdensivu J, Ala-Kotila P, Lahdenperä P. To Demolish or to Repair? Carbon Footprint Impacts, Life Cycle Costs and Steering Instruments. Publications of the Ministry of the Environment 2021:9. Built Environment [Internet]. 2021. Available from: https://julkaisut.valtioneuvosto.fi/bitstream/handle/10024/162862/YM_2021_9.pdf?sequence=4&isAllowed=y [Accessed: 01 October 2021]

[37] Gruiz K, Meggyes T, Fenyvesi E. Engineering Tools for Environmental Risk Management. London: CRC Press; 2014

[38] Dale G, Dotro G, Srivastava P, Austin D, Hutchinson S, Head P, et al. Education in ecological engineering—A

need whose time has come. Circular Economy and Sustainability. 2021;**1**:333-373. DOI: 10.1007/s43615-021-00067-4

[39] Hilli-Lukkarinen M. Architectural planning in the elevation of buildings—Case analysis in Tampere [MSc thesis]. Tampere, Finland: School of Architecture, Tampere University; 2019

[40] Soikkeli A, Koisio-Kanttila J, Heikkinen M. Korjaa ja korota—malleja ja ide-oitakerrostalojenkorjaamiseen ja lisäkerrostenrakentamiseen (Repair and Upgrade—Models and Ideas for Repairing Apartment Buildings and Building Additional Floors). Oulu, Finland: Faculty of Architecture, University of Oulu; 2015

[41] Karjalainen M, Ilgın HE, Somelar D. Wooden additional floors in old apartment buildings: Perspectives of housing and real estate companies from Finland. Buildings. 2021;**11**(8):316. DOI: 10.3390/buildings11080316

[42] European Parlement. Social Sustainability Concepts and Benchmarks, Policy Department for Economic, Scientific and Quality of Life Policies Directorate-General for Internal Policies [Internet]. Available from: https://www.europarl.europa.eu/RegData/etudes/STUD/2020/648782/IPOL_STU(2020)648782_EN.pdf [Accessed: 01 October 2021]

[43] Boström M. A missing pillar? Challenges in theorising and practising social sustainability: Introduction to the special issue. Sustainability: Science, Practice and Policy. 2021;**8**(1):3-14. DOI: 10.1080/15487733.2012.11908080

[44] Metsäranta L. Residents' opinions about wooden: facade repair, roof extension and urban infill—Resident survey to Pukinmäki district in Helsinki [thesis]. Tampere, Finland: School of Architecture, Tampere University; 2020

[45] Bayne K, Taylor S. Attitudes to the Use of Wood as a Structural Material in Non-residential Building Applications: Opportunities for Growth. Australian Government: Forest and Wood Products Research and Development Corporation; 2006

[46] Mallo MFL, Espinoza O. Awareness, perceptions and willingness to adopt cross-laminated timber by the architecture community in the United States. Journal of Cleaner Production. 2015;**94**:198-210

[47] Xia B, O'Neill T, Zuo J, Skitmore M, Chen Q. Perceived obstacles to multistorey timber frame construction an Australian study. Architectural Science Review. 2014;**57**(3):169-176

[48] Kamari A, Corrao R, Kirkegaard PH. Sustainability focused decision-making in building renovation. International Journal of Sustainable Built Environment. 2017;**6**:330-350

[49] Franzini F, Toivonen R, Toppinen A. Why not wood? Benefits and barriers of wood as a multistory construction material: Perceptions of municipal civil servants from Finland. Buildings. 2018;**8**(11):159. DOI: 10.3390/buildings8110159

[50] Churkina G, Organschi A, Reyer CPO, Ruff A, Vinke K, Liu Z, et al. Buildings as a global carbon sink. Nature Sustainability. 2020;**3**(4)

[51] Hart J, D'Amico B, Pomponi F. Whole-life embodied carbon in multistory buildings steel, concrete and timber structures. Journal of Industrial Ecology. 2021;**25**:403-418

[52] Robati M, Oldfield P, Nezhad AA, Carmichael DG, Kuru A. Carbon value

engineering: A framework for integrating embodied carbon and cost reduction strategies in building design. Building and environment. 2021;**192**:107620

[53] Andrew RM. Global CO_2 emissions from cement production. Earth System Science Data. 2018;**10**:195-217

[54] Gustavsson L, Sathre R. Variability in energy and carbon dioxide balances of wood and concrete building materials. Building and Environment. 2006;**41**:940-951

[55] ZabalzaBribián I, Valero Capilla A, Aranda UA. Life cycle assessment of building materials: Comparative analysis of energy and environmental impacts and evaluation of the eco-efficiency improvement potential. Building and Environment. 2011;**46**:1133-1140

[56] Lippke B, Gustafson R, Venditti R, Steele P, Volk TA, Oneil E, et al. Others comparing life-cycle carbon and energy impacts for biofuel, wood product, and forest management alternatives. Forest Products Journal. 2012;**62**:247

[57] Nässén J, Hedenus F, Karlsson S, Holmberg J. Concrete vs. wood in buildings—An energy system approach. Building and Environment. 2012; **51**:361-369

[58] Guo H, Liu Y, Meng Y, Huang H, Sun C, Shao Y. A comparison of the energy saving and carbon reduction performance between reinforced concrete and cross-laminated timber structures in residential buildings in the severe cold region of China. Sustainability. 2017;**9**:1426

[59] Ganguly I, Pierobon F, Sonne HE. Global warming mitigating role of wood products from Washington state's private forests. Forests. 2020;**11**(2):194. DOI: 10.3390/f11020194

[60] Høibø O, Hansen E, Nybakk E. Building material preferences with a focus on wood in urban housing: Durability and environmental impacts. Canadian Journal of Forest Research. 2015;**45**:1617-1627

[61] Švajlenka J, Kozlovská M. Construction-technical specifics of a prefabricated wood construction system. AD ALTA: Journal of Interdisciplinary and Multidisciplinary Research. 2020;**10**(2):373-376

[62] Gosselin A, Cimon Y, Lehoux N, Blanchet P. Main features of the timber structure building industry business models. Buildings. 2021;**11**:170. DOI: 10.3390/buildings11040170

[63] Miles J, Whitehouse N. Offsite Housing Review; Business. London, UK: Innovation & Skills and the Construction Industry Council; 2013

[64] Li Z, Shen GQ, Alshawi M. Measuring the impact of prefabrication on construction waste reduction: An empirical study in China. Resources, Conservation and Recycling. 2014; **91**:27-39

[65] Shamsai M, Whitlatch E, Sezen H. Economic evaluation of reinforced concrete structures with columns reinforced with prefabricated cage system. Journal of Construction Engineering and Management. 2007;**133**:864-870

[66] Zhang X, Shen L, Wu Y. Green strategy for gaining competitive advantage in housing development: A China study. Journal of Cleaner Production. 2011;**19**:157-167

[67] Aye L, Ngo T, Crawford RH, Gammampilaa R, Mendisa P. Life cycle greenhouse gas emissions and energy analysis of prefabricated reusable

building modules. Energy and Buildings. 2012;**47**:159-168

[68] Yee AA, Eng PEHD. Social and environmental benefits of precast concrete technology. PCI Journal. 2001;**46**:14-19

[69] Chiang YH, Chan EHW, Lok LKL. Prefabrication and barriers to entry—A case study of public housing and institutional buildings in Hong Kong. Habitat International. 2006;**30**:482-499

[70] Jaillon L, Poon CS, Chiang YH. Quantifying the waste reduction potential of using prefabrication in building construction in Hong Kong. Waste Management. 2009;**29**:309-320

[71] Wong F, Tang YT. Comparative embodied carbon analysis of the prefabrication elements compared with in-situ elements in residential building development of Hong Kong. World Academy of Sciences. 2012;**62**:161-166

[72] Kong A, Kang H, He S, Li N, Wang W. Study on the carbon emissions in the whole construction process of prefabricated floor slab. Applied Sciences. 2020;**10**(7):2326. DOI: 10.3390/app10072326

[73] European Commission. Long-Term Renovation Strategy 2020-2050 Finland, Report According to Article 2a of Directive (2010/31/EU) on the Energy Performance of Buildings, as Amended by Directive 2018/844/EU [Internet]. 2020. Available from: https://ec.europa.eu/energy/sites/default/files/documents/fi_2020_ltrs_en.pdf [Accessed: 01 October 2021]

[74] Edwards R, Holland J. What is Qualitative Interviewing? London, UK: Bloomsbury; 2013